D0296145
ST. MARY'S COLLEGE LIBRARY
BT12 6FE
SHELVED AT
910
MIN

REGIONAL GEOGRAPHY

Geography

Editor

PROFESSOR W. G. EAST

Emeritus Professor of Geography in the University of London

REGIONAL GEOGRAPHY

THEORY AND PRACTICE

Roger Minshull
Lecturer in Geography, Bishop Grosseteste College, Lincoln

HUTCHINSON UNIVERSITY LIBRARY
LONDON

HUTCHINSON & CO (*Publishers*) LTD
178–202 Great Portland Street, London W1

London Melbourne Sydney
Auckland Johannesburg Cape Town
and agencies throughout the world

First published 1967
Reprinted 1968 and 1971

Printed in Great Britain by litho on smooth wove paper
by Anchor Press, and bound by Wm. Brendon,
both of Tiptree, Essex

ISBN 0 09 082772 4 (cased)
0 09 082773 2 (paper)

CONTENTS

FIGURES

ACKNOWLEDGMENTS

One purpose of this book is to collect and set down those ideas so often discussed in classrooms, libraries and lecture halls yet which seem to have no traceable sources. Thus it is possible to thank only in a general way many people who originated, developed and helped collect together some of these ideas. In the long process of finding material, sorting, writing and re-writing, help is given by many people in completely different ways, but each may be critical at a certain point. I would like to thank, in this connection, the many librarians and book dealers who seemed to enjoy tracing the more obscure articles and books, for their work was vital, and many went out of their way to help.

More particularly, I am very grateful to Mrs M. M. Tyson for her most accurate typing, and to Mrs K. M. King, B.A., for drawing the maps and diagrams, and making my best efforts seem very crude. Most of all I welcome this opportunity to acknowledge my debt to Mr A. A. Arnold, M.A., for his early encouragement, to Dr J. M. Houston for some rigorous training and for reading the manuscript, and to Professor W. G. East, editor of this series, for giving so much of his time to improve that first lengthy, untidy manuscript, and for his generous help ever since.

R.M.M.

EDITORIAL PREFACE

There is only one region—the surface of the earth—on which mankind finds its home. Yet, although much effort is devoted by geographers towards the study of this diversified environment as a whole, it has long seemed necessary, by the methods of 'special' or 'regional geography', to study its component parts. And although nature abhors lines, geographers might appear to adore them, so busily do they engage themselves in delimiting on their maps allegedly significant areas called 'regions'. As a result, every student of geography in school, college of education and university has been taught, read books and attempted to answer questions on, regional geography.

It must be admitted that both the teaching and writing of regional geography have been under fire from critics, not least from geographers themselves, who question the methods, objectives and achievements of those who practise it. Without doubt, some of these criticisms are valid: on an unsure philosophical basis some regional geography, as taught and as written, presumes too much and achieves too little. To some it may appear as an *omnium gatherum* of information about particular areas for which is dubiously claimed regional status. This last is no easy matter to substantiate, for there are clearly very many different kinds of regions of differing scale and rank, conceived on the basis of specified criteria and for a variety of purposes. It is perhaps not fully realised, especially by the younger students, that regions are often conceptual—creations of the geographer's mind—rather than intrinsic and evident realities of the landscape. Some critics, forgetful of a number of regional studies of acknowledged excellence, seek to shrug off this traditional charge on geographers' energies by recommending, as more intellectually rewarding, the study of 'systematic' or 'topical' geography, regional problems, and particular areas by way of selected themes.

Mr Roger Minshull, the author of this book, was himself 'exposed' to regional geography as a student at the University of Oxford and has for some years been required to teach it. To meet his own difficulties, doubts and needs, he has grappled with the study of its methods, concepts and objectives. He has been at pains to discover the method and content of textbooks in current use—a revealing exercise. He has succeeded in clarifying much that has confused many, has underlined shortcomings and pinpointed exaggerated claims. Indeed, he succeeds in showing both the limitations and possibilities of regional geography as a major part of the geographer's craft. In addition, having shown why 'regionalism' stands apart from academic geography, he recalls and reviews Derwent Whittlesey's original contribution to regional theory in respect of the 'compage' which, it might appear, has not received the attention it deserves. By reviving this obsolete though appropriate term as a substitute for 'region', Whittlesey outlined a distinctive method by means of which regional geography might more effectively develop. Not only the younger student, who may well be confused, but also the graduate student, seeking to develop new methods, of the quantitative kind maybe, will be grateful to Mr Minshull for the guidance which he here offers to clearer thinking and better understanding. When, as long ago as 1887, Sir Halford Mackinder asserted that 'it is especially the character of geography that it traces the influence of locality, that is, of environment varying locally', he focused attention on the central purpose of geography. For Whittlesey, too, regional geography was the keystone of the geographic arch.

To this end, namely the understanding of the wholeness of country by way of its distinguishable parts and as environments of which we are part, conceptual regions can greatly contribute as tools. For C. C. Carter, regions were 'good servants, bad masters'; for Professor Spate they are 'isolates for study'. Not that the region provides an easy way to understanding: there indeed is the rub. As the French school of *la géographie humaine* exemplified long ago, regional geography is a highly exacting task, calling for much scientific knowledge and cartographic skill and even more literary art. There is no suggestion in this book, nor should there be, that the writing and teaching of regional geography should be either rigidly conceived or uniformly presented. Rather, as the study of geography increasingly appeals as a liberal education and increasingly affirms the value of its applications, the way lies open to new methods and to experimentation. And success in this challenging task will demonstrate, beyond all else, what geographers can do better than other scholars and scientists.

W.G.E

I

THE REGIONAL METHOD OF DESCRIPTION

So much geography is written on a regional basis that the idea of the region and the regional method is as familiar and as accepted as is Mercator's map in an atlas. Yet as with so many other familiar ideas which we use every day and take for granted, the concept of the region floats away when one tries to grasp it, and disappears when one looks directly at it and tries to focus. Like the ideas of Time and Art it is so obvious yet so difficult to define, but equally so useful, so much a part of geography, that definition would be valuable in many ways. Many short, attractive definitions have been formulated, showing that many geographers feel the need for a short, hard and fast rule which will be their support and comfort on every occasion.

There is also a wealth of conflicting opinion, with detailed expositions of the region; methods of finding, defining, mapping and describing the region, all of which, perhaps, can not be reconciled into one philosophy. At the same time, much of it is valuable and makes up that source of half-remembered ideas where geographers and students of geography first acquired their own very personal concepts of the region. Because much of this is valuable, and often so vague as to seem instinctive, this attempt has been made to bring all these ideas about the regional method together.

Among the many ideas about regions, the two most difficult to resolve are, on the one hand, that the regional method is a means to an end, and on the other, that the discovery and description of regions is the end purpose of geography. Thus we have the two ideas at once, that the region is simply an idea, a method to help in the study of the world; or, in contrast, that regions actually exist, that from an aeroplane a person may actually look at one. The fascination lies not in the question whether one or other of these ideas is correct, but in the fact that they both contain part of the truth, and the answer is a subtle combination of these and other ideas. We might try the historical approach, accepting the former idea that the aim of geography has been to describe the world and the regional method has been the means to this end; but one does not read geography or study the world for long before realising that parts of the earth's surface are so contrasted that it is self-evident that some regions with unique characteristics do exist. Once this exciting discovery had been made the aim of geography could well have been the anxiety to find as many different regions as possible and describe them all.

While a few geographies reveal this attitude (and it must be borne in mind), most descriptive geographies currently in use in fact reveal the former attitude, and in order to avoid any argument at this point about the aim of geography, Hartshorne's carefully considered conclusion will be adopted, that geography is the study of the interrelations of phenomena on the earth's surface and their areal differentiation, to give comprehension of the world. 'Areal differentiation' brings in regional differences where they exist, but the aim is world comprehension, not region hunting.

Given this definition of geography, we know that in fact most geographers have attempted to comprehend that part of the world which each has described, by dividing it into regions. Often no justification is given for the regional division, and it seems to have been taken for granted—this 'instinctive' feeling for regions evident again. While there are a few world geographies, again it is an accepted commonplace rule that the best geographies are of some part of the world less than the whole, the writer having specialised in this area and so knowing it better than most other geographers. At once, to make a start on this world comprehension, we divide the world up to make the task easier. However, even the first division is not simple, or based logically on one criterion. If one tries to make a collection of the best geography books currently available the majority of titles would suggest at least one for each continent, but the collector will end up with such anomalies as North America without Mexico, and Europe with or without Russia as you please. The need for division becomes obvious from the start, but the method of division is the problem.

Anyone who starts to study or describe part of the world has some ready-made units which in many cases serve quite well to break up the whole into manageable portions. Continents and countries are the obvious examples, as coastlines and political boundaries are so precise and reassuring on the map. If these units are too large, states, provinces and counties within the countries may do quite well because when the object is simply to describe part of the world a ready-made and well-known framework is the ideal. There are many excellent books describing Europe or parts of Europe and all but one take the countries as their main units or 'regions'. The contrasts between Spain, Holland, Switzerland and Finland may justify this very convenient method, but should one side of the Pyrenees be described at a different time from the other? Should the French and Swiss Jura be separated? Is north-east France completely different and separate from Belgium? Only Unstead made a systematic effort to treat Europe in a different way, by taking the major relief features as his regions, however much they cut across political units.

There are fewer descriptions of South America, but again the tendency is to take the countries as the main regions: yet some are too large and varied for generalisation to have any meaning and so they are subdivided. Political units begin to fail at this stage and so it seems most convenient to divide Brazil into the Amazon Basin, the Highlands and the South-east Coast; Argentina into the Grand Chaco, the Andes Slopes, the Pampas, and Patagonia; Chile into South, Centre and North, and so on. By this stage there is no common factor except perhaps convenience. Relief, climate, vegetation and settlement have all helped to divide the area for the purpose of description. No criticism is implied here, for, as yet, no rule for dividing continents into units convenient for accurate description has been suggested. The very nature of the task seems to make this inconsistency seem inevitable and it can be argued that as the world and each small part of it is unique, each area to be described must be approached afresh and a convenient basis for division in one continent may be useless in another. The choice of countries and states as regions is seen to be useless in the case of Colombia, Ecuador and Peru. Describing each in turn leads to much unnecessary repetition as all three sit astride the Andes and share the narrow coastal strip, the high plateaux, and the edge of the Amazon Basin. There are differences between the countries, but the most convenient way of treating them is by taking the three relief regions in turn and noting the national variations within them. This may be understood more clearly in the case of the United States. In no case is one of the States (some of which are larger than European countries) a convenient unit for description, and even groups of States have not in fact been much use to geographers because the geometrical boundaries too often have no connection with the relief, agriculture, climate and other features which make one part of the country so different from another that it must be described separately. In Europe the boundaries may have more relation to such features, but often their irregular courses are simply misleading. An irregular line in geography looks more probable than a straight one, and in fact may follow a river or watershed, but is not by that fact alone a good regional boundary.

So in some cases ready-made units work well enough and the interesting thing is to see how those who are using the regional method as a means to an end fashion their regions where no boundaries are marked on the map; where there are no coasts, boundaries or provinces which will suit their purpose. The completed regional texts seldom give any clues to the authors' methods; whether they start with a whole continent or country and break it down into regions, or start from contrasted centres and work outwards until the features being described have changed sufficiently to require separate treatment. The working definition that a region should consist of the largest area over which it is safe to generalise obviously has not been followed in some cases. Reading repetitious chapters one often wonders just why the writer has divided an area into several regions when one description of the whole would have been sufficient. In a recent work, Dr Mutton's *Central Europe*, the division of the Swiss Midlands illustrates this repetition. At the same time there are equally unsatisfactory descriptions where the chosen regions have been too large and have contained so much variety that great subdivision was necessary. One reason for both these tendencies may be the desire to keep regions roughly similar in size, like pieces in a stained-glass window, none very tiny, and none too large, and in order to keep them all the same size sometimes several pieces of red are all parts of one cloak, while one piece contains all the detailed colours of a face.

Hartshorne[1] has likened the earth divided into regions to a picture mosaic, an approximation to the truth which we must not mistake for the truth itself. If this is so, the geographer is using the region in the way that a natural scientist uses a model. Where reality can not be observed and comprehended directly, then a comprehensible model must be used. Regions may not exist in reality, the

[1] Superior figures refer to end-of-chapter notes.

earth's surface may be a continuum, but in other disciplines when we encounter a continuum we often wish to give character to various points along it, especially for practical application. Thus electromagnetic waves vary continuously from ultra short to ultra long but certain stages have been defined as light, infra-red, ultra violet, radio and so on. Our senses and instruments detect them and react to them in contrasting ways but the basic reality behind our excited senses is much more uniform. Where there is a continuous gradual change, but still undeniable contrast from point to point, such a model is essential for us to grasp reality, but we must keep reminding ourselves that it *is* a model.

This approach exemplifies the nature of this kind of geography. It is much more of an instinctive, acquired art than many geographers care to admit, and certainly on careful examination some regional divisions are rough and ready. However, provided the information conveyed by this means is accurate, provided the description is vivid and the book succeeds in giving the reader as accurate a picture of the land and the people as possible, then the regional method has served its purpose. Those interested in the region as a real object may be horrified by this, but to many it is quite acceptable.

When a writer is interested in one part of the world, and his concern is to describe it, another assumption is usually made. Not only is there the unconscious tendency to try to keep regions similar in size but there is often the assumption that all the area to be described will fit perfectly into a regional framework. The descriptive geographer rarely leaves gaps. He may dismiss the northern Apennines or Abruzzi Highlands briefly, but the set is complete, the whole area is accounted for. There is an attempt to make all the regions fit together perfectly and the analogy of the stained-glass window still holds, as there are no places without glass and the window fits the frame. (For some exceptions to this rule see the list given in the Appendix.) Moreover, in this interesting art certainly there is the feeling that regions should not be too long and narrow. Chile and northern Norway are not attractive areas to handle, for the rounder regions are, the more we feel safe with them.

As the regional geography at present under discussion is a method of working, the content and the aim of the work have a great influence on this type of regional device. The geographer is trying to describe certain selected features on the earth's surface and of necessity has to generalise about large areas. This care to generalise about as large an area as possible, competing with the fundamental fact which gives rise to geography, that parts of the earth's surface are different, leads the geographer instinctively to represent the area in question as a set of formal regions. By a formal region we mean a region throughout which one or more of the phenomena such as rock, rainfall, farming or population is sufficiently uniform to permit generalisation, and sufficiently different from the phenomena in surrounding regions to justify the division; or more usually, make division essential. The idea of the region is often so nebulous, and, because of this, often so personal and peculiar to each geographer that Odum and Moore in their book *American Regionalism*[2] collected forty definitions which together may give the reader some idea of the region but certainly show how the idea has grown, varied, been adapted, distorted, and has always eluded satisfactory definition. Perhaps this is its strength; the difficulties may be annoying but the concept obviously is capable of growth and adaptation as geography progresses. Of those forty definitions by far the largest group defines the region as a formal unit. The region is:

An area within which the combination of environmental and demographic factors have created a homogeneity of economic and social structure.
T. J. Woofter

An area delineated on a basis of general homogeneity of land character and of occupance.
R. S. Platt

An area wherein there has grown up one characteristic human pattern of adjustment to environment. American Society of Planning Officials

A domain where many dissimilar beings, artificially brought together, have subsequently adapted themselves to a common existence.
P. Vidal de la Blache

An area throughout which a particular set of physical conditions will lead to a particular type of economic life. R. E. Dickinson

An area whose physical conditions are homogeneous. W. L. G. Joerg

An area characterised throughout by similar surface features and which is contrasted with neighbouring areas. N. M. Fenneman

A complex of land, water, air, plant, animal and man regarded in their special relationships as together constituting a definite, characteristic portion of the earth's surface.
A. J. Herbertson

A geographic area unified culturally, unified at first economically and later by consensus of thought, education, recreation etc. which distinguishes it from other areas.
K. Young

The concept expressed in these various ways is essential to descriptive geography; such geography could hardly exist without it. Try to imagine writing a description of Britain without the regional method. One outcome might be at best an uninspiring blanket description of an island with a cool-temperate, western-margin climate, once clothed with deciduous forest but now part of a larger hay and dairy belt and engaged mainly in industry, with fifty million people, capital London. The other extreme must be a cataloguing of every feature on the surface, working as steadily as a typewriter from the north-west corner to the south-east corner and, like a machine, making no analysis, even for the sake of easier comprehension, let alone for any original comment. This is the method which seems to come naturally to students when they start to describe landscapes, in spite of all previous instruction. Each hill, each valley, each wood, road, railway, canal, hamlet and village is described ruthlessly and doggedly, just as it comes to the eye of the uninitiated or uninterested observer. The reader may be able to glean a few facts from the description, but rarely will he comprehend the landscape, and never will he have his interest aroused in this way. There are, however, books which deal with fascinating countries written in exactly this way.

Admitting that there are writers and readers content at one extreme to reduce the world to mere outlines, and at the other extreme pedantically to list every feature, we can be concerned with those who are anxious not to get the world quickly labelled, nor exhaustively recorded, but who, having some perception of the variety which is not chaos, strive to achieve a world picture. A description of Britain following this middle way would divide the island, contrasting the highlands, the Pennines and the scarplands without examining every hill and without giving a better account of every railway than even the most ambitious railway timetable. So the regional method develops. In this respect it can be seen as the inevitable result of trying to avoid the extremes just described, but, as will be shown later, there are other concepts besides that of the formal region, and if it were just a matter of taking sections of the country one at a time, functional regions or city-regions might do just as well. However, one of the implicit aims of the descriptive geographer is to contrast the different parts of the country because that contrast aroused his interest, makes the description vivid, gives the reader his picture, and is the whole *raison d'être* of this branch of the subject. So it seems that either the geographer has some preconceived impressions of the regions before he starts, or, as soon as he begins to examine new ground, soon achieves some personal impression of the main contrasting areas, and these are usually based on the relief. In most regional geographies, however the fact is obscured, the regions are in fact regions of relief.

As one stands and stares, the three most obvious constituents of a landscape are the relief, the vegetation and the buildings (where these exist). Unfortunately geographers seldom describe the details of the vegetation and buildings, even where these are significant, so it seems that in the field relief makes the main impression. Similarly, as one browses through an atlas, again three features will be pressed on one's attention: relief, communications and towns. While it is presumed that the geographer preparing a book collects or prepares maps of the rocks, rainfall, temperatures, soils, vegetation, products and population, he is already subconsciously conditioned by the relief. Moreover, none of the phenomena listed above except the rock is divided into tangible, static, clearly defined units as is the relief. There may be steeper gradients in the isohyets and isotherms, contrasts in soil, different densities in production and population sufficient to justify a division, but these features are not observable on the ground. Another drawback is the fact that some are changeable from day to day while others show change over a longer period and so are no basis for a regional division by themselves. Instead they are used to reinforce and supplement the stark simplicity of the relief regions.

This is not to imply that there is any attempt to conceal the true basis of regional division. Simply, by the time the descriptive picture is complete, the frame has been hidden and forgotten; but inevitably there is a temptation to make the components of that picture fit the frame. There is even some justification for this tendency because relief has a profound effect on the climate, not just in the sense of high mountains increasing the rainfall, lowering the temperatures and deflecting the winds, but, as Geiger[3] has shown so fully, the slightest undulations in the ground, different aspects, and even the texture of the surface do affect the temperature, evaporation, incidence of frost, length of time snow lies on the ground, and so on. Relief and microclimate in turn affect the soil profile and all three then have a direct bearing on wild vegetation or possible crops. Much of this goes without saying and the geographer almost instinctively takes advantage of this, at the simplest level contrasting the mountain region of cool, wet climate, immature soils and characteristic vegetation with the warmer, drier lowland covered with deeper soils which is more likely to support crops.

Further evidence of this basic reliance on relief is found in that type of geography in which developed the ideas of determinism. The controversy between determinism and possibilism will not be pursued here, and as yet there is no final word on the matter, but throughout the history of geography a proportion of the descriptive works have concentrated on the effect of the physical environment on Man's life and work. As far as the critics of determinism are concerned, too often these descriptions have seemed to show (without conclusive proof) that Man is completely controlled by his environment, or that it is largely a one-way effect, Man having little control over his environment and making very few permanent changes in it. In very primitive societies there may well be much in this, although within one natural environment such as the African savanna we see that the Africans found it possible to concentrate either on cultivation or on nomadic cattle herding and were not absolutely determined to one rigid way of life. In more complicated societies such as our own where there is a much wider choice of possible ways to live and ways of earning a living there is still a tendency in some modern descriptive geography to take it for granted that the way we *do* live is the only way we *could* live, because of the physical conditions. The writers may no longer set out to show how the physical conditions have affected the human geography, or to show how Man has adapted to his environment, but the traces of that attitude still persist. This is partly the result of even the youngest, anti-deterministic geographers necessarily having been brought up on some such works, but it is also equally inevitably the result of the way we usually work, starting with the structure and relief and building up on that base. Thus the base chosen to a large extent controls the way in which other things are built on it.

This is not meant to be an explanation of how determinism started. At least there were enough instances of societies being controlled in certain ways for determinism to develop and have something on which to work. In the same way, but to a much greater degree, there is abundant evidence that a system of regions based on relief is a valid division not only of the physical environment but of most of the phenomena which usually concern the geographer. This method of division has been used since the eighteenth century when Buache visualised the earth's surface as consisting of river basins separated by mountain chains which provided convenient boundaries. His basic idea persists, with many modifications. One development of the idea is to take a river basin as the essential unit, and outside the vast deserts of the world this can be made to work tolerably well. The watersheds provide acceptable boundaries to the regions, leaving no areas unaccounted for. Moreover, natives of the Paris

Basin, the English scarplands and other small, settled areas of Europe may easily assume that a river basin is co-extensive with one community with at least sufficient uniformity in its basic economic activities to be regarded as a cultural unit. Drainage basins have been suggested for a regional division of the African plateau but one has only to consider such examples as the Nile where the ancient Egyptians were virtually ignorant of the Ethiopians and Bantu leading separate lives on the tributaries of the same river, or the Tennessee where the citizens of separate States co-operated to save their own soil only after much strife and federal intervention, to see that the idea of a regional division by river basins is no more an answer to the problem than any other single method mentioned so far. In some areas it will give a useful regional division, in others it is completely meaningless. Just as the river basin can be too large a unit, in such countries as Britain and Sweden where many short rivers run separately to the sea in close, parallel courses it can be too small, and the problem of grouping arises again.

Another necessary modification of Buache's idea has been to define some mountain ranges as regions in themselves, and not to regard them merely as boundaries or at best as frontier zones. The Pyrenees and Himalayas may be effective barriers sparsely inhabited, but the Jura can not be regarded simply as the frontier between the Saône Valley and the Swiss Midlands, while the Andes of Colombia, Ecuador, Peru and Bolivia must be considered the most important region between the difficult coastal strip and the sparsely inhabited interior lowlands. So the major features of relief have become the basis for much regional division, not in the narrow sense of trying to find convenient basins with separate communities, but in the sense that plains, plateaux, scarplands, valleys and hills all present different environments which people have used in so many different ways. This has proved convenient and successful so often that there has been a tendency to play down the drawbacks and the errors necessarily involved. The geographer who is anxious to describe needs a regional framework as quickly and conveniently as possible, in contrast to the methodologists who, so far, have largely criticised the rough and inaccurate methods of the former but have not been able to provide him with a succinct definition of the region; a method of working which can be applied to any part of the earth; nor an alternative to the region.

Some of the error which this regional method involves could be avoided, however, by a different approach to the same material because much of it results from the way most regional geographers tend to work, starting with structure and relief. We are all influenced by our teachers and the overwhelming mass of work which has gone before, and in geography the main theme of that work has always been nature. From ancient Greece until the early twentieth century geographies have been written which were concerned only with the land, the climate, the flora and the fauna; works which would be classed precisely as physical geography today. During the nineteenth century it was long debated whether Man should be included or not. Determinism, with its close attention to all the features of the physical environment and its main concern in Man only to find out how he has reacted to this environment, has had its influence. With the sheer magnitude of this bias to physical geography behind them, even modern geographers interested mainly in human geography usually follow the well-established pattern of first describing the structure, then the landforms, climate, soils and vegetation before getting on to their prime interest, as if this were the only possible order. This is perfectly acceptable when a balanced description is the aim, but one sees the tendency to describe the physical features separately from, and well before, Man's activities, even where there is a direct connection and the two should be interrelated. Thus this inherited bias, this traditional method of working, helps both further to explain the regional geographer's choice of relief as a basis for division, and leads to distortion and omission when the facts of agriculture, industry, settlement and population are considered after, and secondary to, the facts of nature.

There is some justification for the attitude that a division of the earth's surface into very large major relief regions coincides with changes in the climate, soil and vegetation sufficiently for the descriptive geographer's purpose; but there is much less evidence that there is a similar correlation with such things as farming and the distribution of population. In Cheshire, for example, the low plain on the Triassic rocks may be contrasted with the Pennine upland on the Carboniferous rocks. Boundaries may be drawn where the temperatures and rainfall change more rapidly, where the steep western slopes overlook the plain, but only the most general, over-simplified junior school textbooks could then pretend that this contrast could be extended to asserting that the plain is a well-populated dairy-farming and industrial region while the upland is used for sheep farming. In fact Man has ignored the physical boundaries and the dairy farming extends several miles into the hill country and this zone is more densely populated than the rest. Thus no contrast can be seen in the maps of land use, population, settlement and other purely human features at the same boundary as the contrast in natural features, yet the approach to these facts starting with the natural features gives a very strong temptation to ignore or play down such lack of correlation as long as it is sufficiently small. So the scale of the work is a factor; in a geography of Europe a simple contrast between the Pennines and the lowlands may be acceptable, but in a description of northern England it definitely would not.

When dividing the earth's surface for the purpose of description it is possible to start at both ends of the scale. Herbertson, for example, started with the whole world and divided it into his natural regions, defining as many as were necessary to distinguish the different climates and types of natural vegetation which are found in the world. At the other extreme Unstead built up some regions, extending each as far as possible from the chosen centre as long as its outskirts shared the same characteristics as the centre. Similarly, it is equally possible to start a regional description with the facts of population as with the facts of relief. Paul Vidal de la Blache considered the facts of the distribution of population to be the most important in geography and they formed the starting point for his work. Many geographers honour his name but very few seem to follow his lone example; the habit of centuries is too strong. Admittedly a complete reversal to a method of working which always started with population could lead to as much error as the present method, but there is no likelihood of such a complete reversal and the extra discipline needed to decide which phenomenon will provide the best starting point in a particular country might keep one more constantly aware that all the phenomena under consideration do not necessarily correspond in their patterns to the major landforms. Starting from the landforms, perhaps partly influenced by determinism, one has a tendency to state the facts of the physical environment and then to go on to see to what extent Man fits in with them. Starting with a statement of Man's activities and material constructions, one may then go on to a more impartial examination of the physical features within the regions defined by human phenomena with much less possibility of making the facts of relief, climate and soil fit into any patterns which have already been discovered in the human phenomena.

While this method might lead to less error and less generalisation it would certainly make the task doubly difficult. The geographer would be faced with two sets of problems instead of one; with the usual method he can confidently start with structure and relief, and his main problem is to find a framework for regional division. With the method suggested above there would seem to be infinite possibility in the choice of a topic with which to start, without any easing of the problem of regional division. By this stage the problems are becoming more those which interest the geographers who believe that regions actually exist and can be defined with some accuracy; problems with which the geographer who sees regions only as the result of a method to divide up his area for description is impatient.

Thus the regional method in its most familiar form is to divide the world into formal regions using the most convenient units, which are either already established or easily perceptible. One region is contrasted with another, and wherever possible the regional unit is the continent, island, country, state, county or province which can be marked definitely on the map. In default of this, the obvious major landforms provide convenient units for regions which are essentially relief regions. Moreover, as the regions described by this method tend to be formal, homogeneous areas, this method inevitably leads to the danger of thinking of each region as completely different and separate from its neighbours. Only when we begin to consider the possibility that regions exist outside the geographer's mind can we conceive of functional regions with very complex interconnections. Although a geographer concerned only to describe the earth's surface as accurately as he can need not be at all perturbed if regions are shown to be products of his imagination, providing the regional method is useful to him, consideration of the types of regions which have been defined will at least extend the scope and usefulness of that method.

[1] Hartshorne, R., 'The Nature of Geography', *Annals of the Association of American Geographers*, 1939, ch. 9

[2] Odum, H. W., and Moore, H. E., *American Regionalism*, New York, Henry Holt & Co., 1938

[3] Geiger, R., *The Climate Near the Ground*, OUP, 1957

2

REGIONS AS REAL OBJECTS

Regions are genuine entities, each of which expresses both natural and cultural differentiation from its neighbours. G. T. Renner

Every region has its unique character to which contribute the features of soil, atmosphere, plants and man. R. E. Dickinson

It is this tendency of all tribal traits of culture to coincide with economic traits that gives a regional character to culture as a whole. C. Wissler

Between the continent and the village is an area sometimes larger, sometimes smaller than the state. It is the human region. Lewis Mumford

Many planning officials, ethnologists, sociologists and other students of Man in his environment, as well as many geographers, regard regions as facts of the earth's surface. For them a division into regions is not a means to the end of describing that surface: the end is to define and describe the regions which actually exist. This necessitates a much more accurate concept of the region and a much more consistent attitude towards it than does the approach of the descriptive geographer. If one asserts that regions exist, one should be able to map and describe them with some precision, using standardised methods. While descriptive geographers, as has been shown, have some grounds for using the regional method without insisting on the independent existence of regions, the regionalists have only the same facts to work on with exactly the same shortcomings. Therefore, their much more definite case for the existence of regions is seriously weakened by the same arguments which form a milder attack on regional descriptive geography. However, the types of regions which have been observed, defined and described are so varied that they will need some detailed examination. Some will be found to be valid, but others to be just as vague as the concepts of the descriptive geographers.

When the regional concept is simply a tool, the interest lies in the complexity of the features of the earth's surface in one area. But those who study regions for their own sake do not necessarily concern themselves with the full complex. Thus at one end of the scale of types of regions we have the single-feature region. This is the easiest to define and the most acceptable to those who are not convinced of the existence of regions. The world may be divided into relief regions, each defined by a certain contour or change of slope, simply for their own sake, with no suggestion that these regions correspond with the regions of any other single feature. Similarly, on a world scale, soil and vegetation regions may be defined. Podzols, chernozems, lateritic soils and so on are sufficiently different to justify this division but immediately we come up against the problem, already mentioned, of drawing exact boundaries. The lines on a world map may be exact enough at that scale, but when dealing with small areas one realises that soils change gradually from one characteristic type to another and more often than not the boundary of a soil region must be arbitrary. A world map of natural vegetation regions may mark off Jungle, Savanna, Semi-Desert and Desert quite definitely, but this does not correspond with the infinite gradation on the ground.

With farming regions such problems may be overcome. Here Man is a factor, with his tidy mind, so that the edge of a region may well coincide with a fence on one side of which is a cultivated field and on the other side virgin forest or open rough pasture of the moorland. Within the cultivated area, certain criteria can be chosen and regions defined accordingly. One method is to map the area where a certain crop-combination is predominant, say the wheat, barley, roots and clover rotation of East Anglia or the wheat, vines and olives of the Mediterranean area. Certain crop and animal combinations are found to predominate in different areas, chosen by the farmers with regard to soil, climate, demand and local custom, and the region where one combination is characteristic can be divided from and contrasted with the surrounding regions. A simpler and more precise method is to concentrate on one crop or one animal and to map the yield per acre or density per square mile. This gives maps similar to those in the Oxford Economic Atlases where the dots illustrating the distribution of one commodity justify the recognition of such regions as the Corn Belt and regions which concentrate on one or two products as in New Zealand. With a map such as that showing the distribution of maize the central USA contrasts so strongly with the rest of the continent that a boundary may be drawn with little difficulty. Where the contrasts are not so great over a small distance a certain minimum yield may be taken as the boundary and provided this criterion is made clear then the regional division is acceptable. Taking each commodity by itself gives single-feature regions, but farming regions are combinations of these, with the necessary extra complexity and difficulties of definition. While they can be defined satisfactorily, they have the fault that emphasis tends to be placed on the characteristic crops and animals, especially those for sale and export, while other crops essential to a sound crop rotation and perhaps destined to be used on the farm to feed animals are played down or ignored.

Study of large-scale land-use maps will show how valid farming regions are. Each field may be marked as having a different use from its neighbours, but the wider view will show some areas where crops or orchards are much more common than permanent grass for dairy cattle. Yet it is the geographer who divides this untidy mosaic into neat regions. There is enough contrast between separate areas to give the idea that regions exist but the work of defining and describing them greatly strengthens this idea. It is a common complaint that the very fact of dividing a country into regions simply for convenience of description brings those regions into existence. Once the first geographer to describe an area has used this method, his regions, however arbitrarily chosen, become established and virtually sacrosanct. In full descriptive geography this original division at least saves later writers the trouble of making their own framework but in the case of single-feature regions this may lead to distortion in the very place where the regionalists would seem to be most exact. Once relief regions, or vegetation regions, or farming regions have been defined it is just that much harder to take an unbiased view of the facts. If the original work was accurate and dealt with static phenomena this does not matter greatly; but if the original work was inaccurate or dealt with constantly changing phenomena its influence on later work may be a serious drawback. The Cotton Belt of the USA in all probability was well defined originally. Since then cotton production has almost died out in the east, Texas has become much more important with vastly different modern methods, while production has spread into New Mexico, Arizona and California, well beyond the western boundary of the belt (which was based on the rainfall), where cotton is now grown under irrigation. It is possible to re-define the Cotton Belt in a new place, but at present the tendency is to cling to the old region with the correction that the east is in decline and to regard the areas of production in the western States as 'outliers' of the established Cotton Belt.

Whether these single-feature regions do exist outside their observers' minds or not, they have become as useful to geographers in their own way as the regional method considered in the first chapter. If we admit that at least the regional method is useful in enabling the geographer to examine and describe all the features of the earth's surface in one limited area which is unique, then the process of examining the distribution of one feature throughout the world is useful in correcting, balancing and complementing this concentrated work in depth at one place. Although regions are under consideration in each case, here we have, of course, the two original and essential branches of geography, Special Geography and General Geography. Here one must be very careful of terms. Special Geography, the branch which is concerned with the unique characteristics of the earth's surface which result in each area from the combination of the complete physical and human environment, is usually called regional geography. Strictly speaking, it is complete descriptive geography on a regional basis. General Geography, which is contrasted with it, is concerned not with the sum total of phenomena at one place, but with the distribution and variations of one phenomenon all over the globe. But this type of study has led also to a regional division, not so much from a need to divide up the area for convenient handling but rather because the variations of the phenomenon themselves suggest this division. This is the basic argument for the actual existence of regions and in the case of each isolated feature seems valid. As has been suggested, the validity of a simple division of the world into relief regions has given much strength to the claims that the regional method is more than a device and is in fact a description of actual individual regions.

So General Geography may have as much right to be called regional as has Special Geography, but the regions in question are of a different type. Each relief region, vegetation region or farming region has its own characteristic relief, vegetation or farming which is different from that in adjoining regions. This type of single-feature region is again of the formal type mentioned earlier; it is homogeneous within the limits set by the observer, but while the region must be different from adjoining neighbours, it is not necessarily unique.

Special Geography does describe unique regions. There are no two areas exactly alike in every respect on the face of this planet. Even if everything seemed to be identical, one region would have one quality absolutely its own—its position relative to the other—just as one historical period can never be like another if only for the fact that it comes before or after it. But no two regions remotely approach this condition. The Congo Basin and Amazon Basin may both be tropical jungles but with slightly different flora and fauna, and vastly different histories. Spain, Southern California and Swanland may have similar climates but it would be hard to mistake one for the other. This uniqueness of each region is the mainspring of geography which in fact makes this subject unique. It can never claim to be just another science because the objects it studies are specific and so geography in its complete form is idiographic.

However, the branch of geography called General or Systematic Geography has revealed generic types of regions, is nomothetic, and may therefore be considered a natural science. But while descriptions of special regions can form a satisfying study by themselves, Systematic Geography by itself can become an abstract, incomplete, unsatisfying study. It dissects the world into its component parts and describes them in an isolation which is completely unnatural. A study of world landforms considers in addition only the climatic factors which have been the agents of erosion; a study of crop production again only those physical and economic factors which decide whether it is grown or not. One can imagine that the geomorphologist would be delighted if cities, levees, sea walls, groynes and every other structure could be removed for him to study the landforms without hindrance. The economist finds it difficult to account for other than strictly economic motives in, say, the location of industry. To him, the location of a factory should depend solely on the land, capital, raw materials, labour, communications and markets available, not on the fact that the inventor happened to live in Cowley or that the managing director's wife wanted to move nearer to London.

This scientific approach, searching for generic types, has been successful in the case of single-feature or limited-feature regions, especially with non-human phenomena, giving what is by now a familiar world pattern, but it must be stressed that while these regions can be defined, and may in fact exist, they are only part of the phenomena on the earth's surface and do not exist in isolation. In the human body we may conceive the separate entities of the skeleton, the blood vessels or the nervous system. These can be defined and do exist, but not independently of each other and not in the sense that they can be observed directly in real life. The coloured diagrams of the body's organs may give a useful concept of the reality, but just one of the shocks of watching an operation is to find that real organs are all the same colour and packed together in a shapeless mass. So on the earth's surface one may perceive vegetation regions but they are not like the parts of a botanical garden; each is also the site for farming, industry, settlement and other features which may cut across its boundaries.

Certain features have been isolated and we accept the maps of landform, climate, soil and vegetation regions as representations of reality just as we accept diagrams of the skeleton or alimentary canal. Moreover, maps of these features reveal similar regions in significant parts of the world and a meaningful pattern has long been recognised. In each continent lateritic soils are found astride the equator, the black earths just outside the tropics, then the podzols stretching to the arctic circle, suggesting at the very least that latitude is a factor in their development. The pattern of vegetation is probably most familiar, with the jungles astride the equator and the size and contiguity of the plants decreasing polewards to the tropical deserts. Beyond these are the five regions of evergreen xerophytic woodland, one on the western edge of each continent, then the great continental grasslands, the coniferous forests and the tundras. Here, more clearly, position on the continent is seen to be as important as latitude. Certain regions recur on the west coasts, or in the interior of each continent. The variety of climatic maps is much greater than that of soil or vegetation maps, but a similar pattern can usually be seen. There is an interesting essential difference between soil and vegetation regions on one hand and climatic regions on the other. The facts of the former pair are mapped in each case, and immediately one has a map of the regions. But a map of annual rainfall or of seasonal temperatures is not a map of climatic regions. Strictly speaking, climatic regions are multiple-feature regions involving at least temperatures, rainfall and winds and sometimes including such features as evaporation or evapo-transpiration. The explanation, however, has more to it than the realisation that climate consists of more than one element. There is also the attitude that maps of these elements do not by themselves make clear the differences, say, between the climates of Britain, Sicily, Ceylon and Japan. The coloured maps of soil and vegetation are certainly generalised and idealised, but not to the extent that a map of climatic regions is an idealisation of the actual climates. In fact the maps prepared by such climatologists as Supan, Kendrew and Thornthwaite are something different from the distribution maps found in the atlas. It is difficult to draw definite boundaries for soils or vegetation but the elements of climate are perfectly continuous and differ from place to place only in degree, not in kind. Yet definite boundaries are drawn and the regions are defined and classified into generic types.

This is another example of the objective observation being combined with the subjective rationalisation to establish regions. The essence of the region does exist; it is not an idea without foundation, but the end product is something greater and more definite, simpler and easier to comprehend than chaotic reality. At certain places on the continents, one can observe, measure and record characteristic climates, so that it is possible to contrast places with rain all year and a small temperature range; places with rain in winter; places with rain in summer and hardly any temperature range; places in the hearts of the northern continents with a very wide temperature range between summer and winter. Having chosen such centres on the basis of greatest contrasts, lines are drawn midway between them, boldly marking off the Cool Temperate Western Margin climatic region from the Warm Temperate Western Margin, and so on. This seems to have been the method with the older classifications and as the rainfall and temperatures changed gradually, continuously, from one weather station to another, only the centres of the regions had the typical characteristics as described. Moving from the edge of one region into the edge of the next would reveal less difference than moving from the edge of one region to its centre.

Another method of arriving at climatic regions, at least in the way the work was finally presented to the general public, was to build up a chessboard from the separate elements. First, the temperatures, decreasing polewards from the tropics, gave zones or bands round the earth parallel to the equator, distinguishing the Hot, Warm, Cool and Cold zones—the climatic regions of the Greeks. This divides the chessboard into bands from side to side. Secondly, rainfall is observed to decrease inland and, as a result of the distribution of the continents, in the majority of cases this is in an east and west direction, very roughly at right angles to the temperature zones. Thus we arrive at the Western Margin, Continental and Eastern Margin zones. These provide the divisions up and down the chessboard dividing it into squares or regions. A. A. Miller[1] constructed a diagram of an ideal continent divided by this method and his world map shows its application to each of the actual continents. This completely different approach to the problem can produce or reveal climatic regions similar to those of the first-mentioned method. But the boundaries are no more valid and do not often correspond with any marked sudden change in the climate from one place to another.

More recently Thornthwaite has attempted to find significant boundaries for climatic regions, demonstrating a third approach to the delineation of regions. Most divisions of the continuum of rainfall or of temperature are arbitrary, especially when Man is considering the climate in relation to himself, but certain critical limits have been found. There can be too little rainfall for commercial arable farming. Twenty inches was for some time regarded as the minimum, but obviously this depends on evaporation, run-off, moisture retention of the soil and the requirements of the crop. As new varieties of wheat have been bred the limit in some parts of Canada and Australia has been reduced to fifteen inches. However, as rainfall decreases, as a general rule, it becomes less reliable, so that below a certain figure in a given area irrigation becomes necessary for secure farming. In many classifications ten inches of rainfall marks the edge of the true hot desert. The 'certain figure in a given area' can be determined and can form a worthwhile boundary of a climatic region. Throughout the scale each crop requires rainfall within certain limits up to those for tea and rubber which require some of the highest amounts of all. Similarly minimum temperatures are critical. The isotherm marking the edge of the area with at least 200 frost-free days was long recognised as the northern limit of the Cotton Belt in the USA. Most types of wheat need a temperature of at least 43° Fahrenheit in order to germinate, and similar temperature controls exist for all other crops and vegetation including the deciduous and coniferous forests which are sometimes considered as major natural regions.

Thornthwaite tried to take these obvious facts much further and to base his climatic regions on exact measurements of the combined effect of the elements on crops and vegetation using grass as his control. This work has involved very careful measurement, not only of rainfall and temperatures but also of potential evapo-transpiration; the constant realisation that the effect of each element of the climate varies with variations in the other elements; and a search for boundaries which cause changes in the crops and vegetation and therefore coincide with them. Thornthwaite's work has become increasingly detailed and complicated.[2] It has been criticised as too complicated and not entirely successful, but it points the way to this type of regional delineation. Even where regional *differences* seem so marked and so significant, with intangible phenomena such as temperatures and evaporation it is exceedingly difficult to define an *area* which is different from another. As Thornthwaite's critics have said, perhaps in some cases geographers are wasting far too much time attempting the impossible and undesirable. Whatever the method and whatever the aim, all these, and other, classifications of climate show a similar basic pattern, and similar types of regions in the same latitudes north and south of the equator and similar distances from the coast.

The single-feature regions just considered are often combined, with some justification. Soil and vegetation may be observed readily and independently but this is not true of climate. Weather stations all over the world have had to measure and make records over at least thirty years before the facts of climate could be pieced together. Because of the observed control of climate over vegetation, and these two in turn over the soil, soil and particularly the vegetation have often been used as the data for the delineation of climatic regions. The resultant vegetation has been regarded as a readily observable indication of the climate which requires much more painstaking and patient observation. Early classifications such as Köppen's were actually based on, or adapted to, vegetation zones and although since that time it has been made clear and emphasised by many field workers that in huge areas the 'natural' vegetation shown on the atlas maps does not exist any more, we still cling to these ideas. Not only have forests and grasslands been completely replaced by crops and pastures in much of North America, Asia, Australia and Europe, but in other parts of the world where there is wild vegetation this has often been proved to be secondary growth since Man has destroyed the virgin cover and the present plants hiding his clearings are not the same type, density and height of the original. Much of the Congo jungle is secondary growth where the Africans have moved over it with their slash-and-burn shifting cultivation. In the East, areas cleared of forest often become bamboo thickets when they are abandoned; they are colonised by other plants which gain a hold as the result of human interference and the climax vegetation, similar to the original, may never be achieved.

In spite of this, soil, vegetation and climatic regions are still combined to form natural regions. The causal connections between the three are considered more significant than the instances where a certain type of climate, vegetation and soil are not co-extensive. With the tendency on one hand for some geographers, especially teachers and popularisers, to want these three types of regions to fit together, and on the other hand for the research workers to base one type of region on phenomena belonging to another, the validity of single-feature regions is extended to very complex regions involving all the physical features of the earth's surface. Even the relief may be included by playing down the accidents of geology and earth movements, and stressing the influence of altitude and mountain barriers on the climate, and recognising landform regions where one of the agents of erosion, water, ice or wind has been dominant, producing recurring characteristic landforms. Since Herbertson combined the work on single-feature regions of physical phenomena and not only publicised the results but made them an important part of the teaching of geography in Britain, both at Oxford and in the schools, these natural regions have become established as accepted facts.

Although Herbertson did not pay much attention to the human geography of these regions, others went on to show the coincidence of human differences with these differences in the natural habitat. Each natural region was said to be the location of a certain definite way of life. For example the ice cap and Tundra of the Arctic is the home of the Eskimos; the steppes are the home of the Tartar horsemen; the deserts of the wandering Arabs and most primitive Bushmen; the savanna of the hunters and the jungle of the shifting cultivators. This attempt to link one typical way of life to each natural region, an optimistic attempt to get the whole world neatly pigeon-holed with complete regions, was often so crude that it detracted from the value of the natural regions themselves. The worst examples show Man completely controlled by the physical conditions of the region, the implication being that as he is so controlled, the way of life which exists is the only one which could possibly exist in that region and therefore that way of life is an essential part of that region. Each natural region was shown to have a rhythm, largely imposed by the climate, with which Man has to harmonise if he is to live there. This was achieved largely by selection of the information about life in different parts of the world. It seems not that there was deliberate distortion or suppression of facts so much as a complete disregard of whole societies in the blind faith that one society was completely typical and characteristic of each region. This tendency was strengthened by lack of information when many exceptions to the chosen norm were just unknown. Herbertson's work at the turn of the century is seen as an attempt to coordinate and present all the information which had become available about the world as a result of the exploration and research in the nineteenth century. While there may have been just enough information in Herbertson's time for him to delineate and describe the natural regions there certainly was not enough information for these additions to his work by others.

The accumulation of knowledge since then has revealed that there is not just one possible way of life in each natural region, nor an annual rhythm which completely dominates Man. The work of countless ethnologists and anthropologists this century has shown up the weakness of this world system of regions, but it is precisely this seemingly logical correlation of natural environment with human activities that tends to be made in more localised regional geography. Yet time and again extra, detailed research into all the features of human activities shows that the equation of one formal natural region with one formal region of human phenomena is a very crude over-simplification. As an example of this tendency to over-simplify and to take one group of people as characteristic of a whole region we may consider the African grasslands or the savanna. Throughout this huge, comparatively level tableland with its wet and dry seasons as the convectional rains follow the overhead sun the fascinating variety of African peoples have been labelled collectively 'Bantu', and shifting cultivation based on maize and millets is usually given as the typical way of life. Sometimes a certain tribe, usually the Masai, has been described as an interesting exception where the people are hunters and herders and regard cultivators as slaves. Yet within the region there are at least the following distinct peoples: MOHAMMEDANS, Fulani, Hausa, Mandingo; NILOTES, Shilluk, Anuak, Luo, Lango, Acholi, Nuer, Dinka; NILO-HAMITES, Bari, Lotuka, Didinga, Beir, Masai, Turkana, Nandi, Suk, Karamojong, Iteso; NORTH-EASTERN BANTU, Akikuyu, Wachagga, Akamba, Wanderobo; LACUSTRIAN, Banjoro, Banyankole, Baganda; EASTERN BANTU, Swahili, Awankonde, Wafipa, Wabena, Wayao, Anyanja, SOUTHERN BANTU, Barotse, Shona, Zulu, Matabele, Swazi, Bathonga, Bechuana, Basuto, Bavenda, Ovaherero, Ovambo and Ovimbundu, each with sufficient differences to distinguish them from the rest. All ways of life are represented; gathering wild foods, fishing in the lakes, hunting the big game, herding cattle, and growing crops. A one-volume geography of Africa could not deal with all these peoples, while the several-volume works which do deal in detail with each of these peoples are of limited interest, mainly to ethnologists; but the best regional geography must strike a balance between over-simple generalisation and unnecessary detail. Too often detail has been ignored for the sake of neat regional correlation and to try to establish a pattern of generic human regions corresponding to the natural regions. The idea, based on scanty early information that the great cool grasslands are the homes of herders, the tropical grasslands of cultivators and so on, has not been borne out. The Tundra has miners, Eskimo hunters and Lapp herders; semi-deserts had the Paiute food gatherers, Comanche bison hunters, Kazak herders and cultivators, and the Hopi and Yuma flood farmers, while the Equatorial forests have Pygmy gatherers, Bantu cultivators and Negrito hunters. The different origins and tribal customs of these peoples are seen to be just as important in shaping their daily lives and the way they make a living as the climate. The geography of white settlers in the Americas, Africa and Australia, together with Chinese and Indian merchants, imported African, Indian and Chinese labour, shows both that origins and traditions can control the way of life as much as the climate and that in any one natural region all types of society from the simplest to the most complicated, from food gatherers to machine minders, can be found and are not to be explained away too simply.

[1] Miller, A. A., *Climatology*, Methuen, 1931

[2] See also Hare, F. K., *Geography*, April 1966

3

FORMAL AND FUNCTIONAL REGIONS

The use of the regional method to describe a country implies that the region is regarded as a formal unit, but the description of regions in their own right is not limited in this way and it is possible to recognise functional regions. The formal region has been regarded, up to this point, as a homogeneous unit, but even this requires some qualification. The idea of the formal region as a method of description is that whatever is stated about one part of it is true of any other part; it is the largest area over which a generalisation remains valid. This is simple enough with single-feature regions; a region can be of limestone, flat, be grazing country from one end to the other, or have a hundred people to the square mile, but with multi-feature regions very rarely is each feature completely uniform throughout. However, often enough it is found that while there is great diversity over a small area, repetitions on the same pattern are found in the larger area, and to save unnecessary repetition by describing each repeat in the pattern, one generalises. In south-east England, describing a small area from the south coast across the South Downs into the Vale of Sussex, one finds great diversity in the physical environment and the land use. Taking the wider view, the strata of clays, sandstones, chalk and limestone are seen to outcrop again and again forming a series of escarpments and vales through the North Downs and Chilterns to the Cotswolds. Local microclimates and soils show a similar repetitive pattern matching the rocks and relief, and in turn having some effect on farming and settlement so that the four ranges of hills have much in common in their crops and stone houses although separated by the wide clay vales where market gardening, dairy farming and half-timbered houses are much more characteristic. Thus at a certain level of generalisation this may be labelled the region of the scarplands, or the scarps and vales, with sufficient internal uniformity of recurring features to distinguish it from East Anglia, the West Midlands and south-west England. This is begging the question, raised earlier, whether all the features described in geography do coincide into neat regions, but it demonstrates the ideas of the homogeneous and formal regions which do not necessitate complete uniformity and monotony (see Fig. 1).

This grouping of small homogeneous regions such as a limestone plateau and a clay vale into formal regions such as the scarplands reduces one of the major problems of regional geography—where to draw the boundaries; and leads to the idea of functional regions. Consider a straightforward description of north-west England using the regional method with its heavy dependence on relief. A writer trying to generalise over as large an area as possible and therefore dividing his area into formal regions must draw one boundary somewhere along the edge of the Pennines. Mapping the Triassic rocks of the Lancashire and Cheshire plain and the Carboniferous rocks of the Pennines and Rossendale will give a slightly different boundary from the line marking the change of slope from the plain to the hills. Isopleths for a certain temperature or rainfall may coincide with one of these but again the boundaries of land use and those marking a significant change in population density give slightly different positions. A probable compromise would be to draw the boundary of the lowland region some way up the Pennine slope, and certainly up the valleys of most of the tributaries of the Irwell and Mersey in order to include all the old textile mills strung along them at the now defunct water-power sites, as these mills can be included with the industry of the lowland focused on Manchester and Liverpool. Apart from the small area where the industrial regions of Lancashire and Yorkshire are continuous, between Colne, Keighley, Oldham and Huddersfield, the Peak District to the south and the rest of the Pennines to the north can then be generalised as cold, wet, bleak upland moorlands, largely covered in peat and heather, used mainly for sheep farming, recreation and water supply.

In dividing the North-west thus, the imaginary writer's purpose has been served, for with the minimum division he has described the essential features of the plain and the hills, contrasting them for emphasis. But one unintended result of this method is to give the reader the impression not only that these two regions are different, but also that they are separate and completely unconnected. This impression that each formal region is completely isolated reaches its most exaggerated form in the treatment of a whole continent by a comparatively large number of small regions. Inevitably, however one works through the chosen regions, some pairs which are adjacent must be described in widely different parts of the book. This problem is seen in Spate's *India and Pakistan*[1] where he has many really detailed sub-regions, and in James's *Latin America*[2] where the regional treatment of each country is further separated by much general geography. Yet several schools of thought would divide northern England in another way. Following the ancient idea of basins the main divide would be down the centre of the Pennines, separating the region draining to the Mersey from that draining to the Humber. This seems more logical in many respects. There is no artificial boundary along the Pennine edge, where none exists in real life, while the new boundary is in the no-man's-land of the moorlands where there is a genuine divide in the land use and distribution of population. The two regions thus delimited are much more satisfying concepts—but each is that much more complex and difficult to describe.

A similar division would be made by the student of functional regions. To him the essential unit is not the homogeneous area over which one can generalise, artificially cut off from its surroundings, but the complex of areas which function together as a whole. In a functional regional division the western Pennines would be included with Lancashire, as they are tributary to the industrial area, supplying limestone for the chemical industry, soft water from the millstone grit for textiles and for millions of people, young stock for the lowland farms, fresh milk, and space for recreation. For the functional region is essentially diverse; it is a place where adjacent contrasting physical environments permit a variety of activities which are complementary in supporting the life of the whole. It is a classic example of unity in diversity; the uniform features of a homogeneous region may just *be* together by chance, but the parts of a functional region *work* together and are to a degree dependent on each other.

Carried to the extreme, a functional region would involve all Man's activities. Le Play's Valley Section, an idea worked out in the second half of the nineteenth century, gives an idealised picture of such a region. High up in the hills round the head of the valley may be hunting, or extensive pastoralism; slightly lower, on the steeper slopes, is the forestry, mining, charcoal-burning and perhaps primary refining of the ores; lower still in the mature stage of the valley

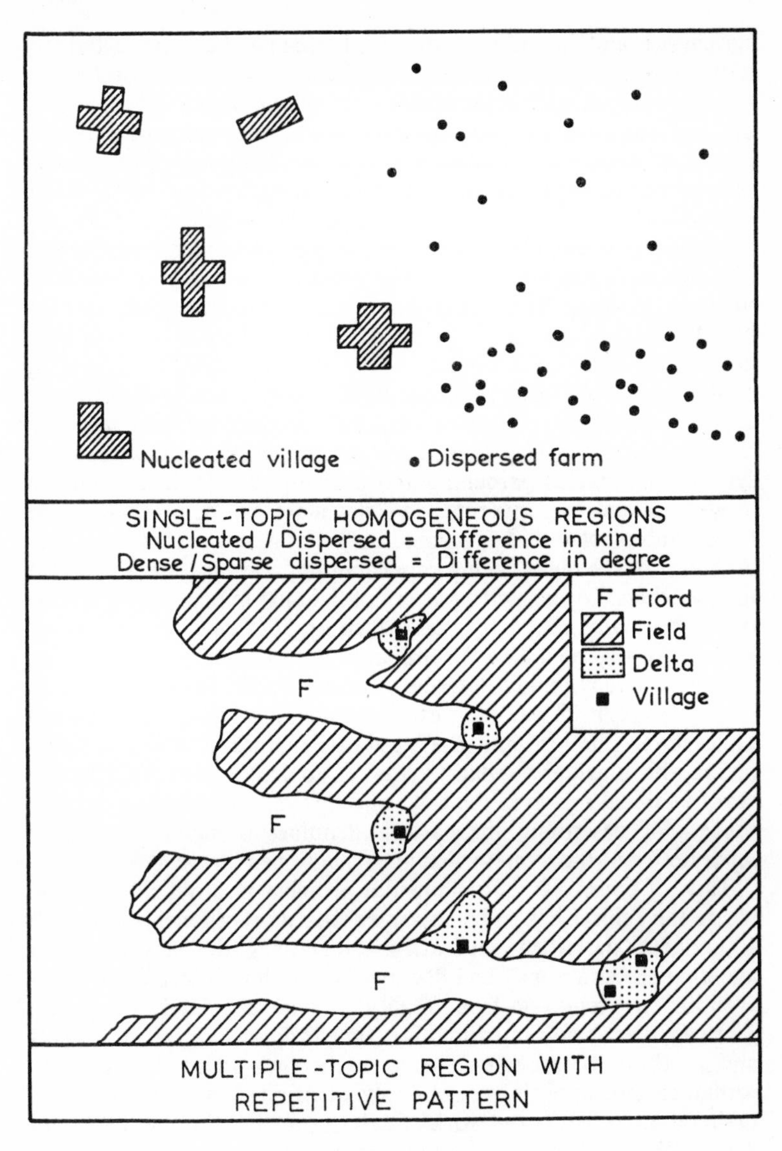

Fig. 1 Formal regions

agriculture and intensive pastoralism become the main activities, settlements are larger and more permanent; finally, in the lowest, widest, warmest and most fertile part of the valley are the most demanding crops, the dairy farming and market gardening for the city which contains the industry, commerce, administration, art and entertainment, which is the focus of the valley and the greatest achievement of the region. Considered in isolation and greatly simplified, it is like the feudal fields, village and manor surrounded by forest or waste which to a very great extent had to provide all the food, clothes, shelter, companionship and religion necessary for life. A very similar idea is embodied in Von Thünen's theory of land use.[3] Again the idealised situation is a city surrounded by a farming region completely isolated from other regions. Furthest from the city is the extensive pastoralism, and successively in concentric circles toward the city ever-increasingly intensive arable farming, sylviculture, market gardening and dairying. Von Thünen confined himself to land use, this time on a flat plain, but again there is the basic concept of the region containing a variety of features and of each part of the region being dependent on, and functionally related to, the others. Given such a pattern of land use in real life it could be treated formally, taken apart and each ring described separately, or it could be regarded as a functional unit and described as a whole. With only one region to describe, the latter method seems obviously better, but consider north-west England again. Having described Lancashire, Cheshire and the western Pennines as one functional region, the imaginary writer must go on to describe the West Riding as a functional region in the east; and this will involve the repetitive process of describing the eastern Pennines in addition to, and separate from, the western Pennines mentioned above.

This, of course, is a fundamental problem in any field of study where the subject has to be separated into its components and each component described separately. Sometimes the advantage of avoiding repetition may be offset by the resultant information being even further removed from reality than is absolutely necessary. The best description must involve some exaggeration of the subject and needs to be concise. Therefore the method of division into formal regions appeals very strongly when one is studying an area, taking it apart before the work of description, but this method is not necessarily the best for the final presentation of the information which has been digested and selected by the geographer. In some cases the most accurate and useful description is that using the method of functional regions to show the interdependence of areas.

Neither method is completely satisfactory; the more usual formal description often lacks the insistence that these homogeneous regions are part of a greater whole, but the functional description, by cutting across formal boundaries, often destroys valid concepts of these homogeneous areas. A pedantic striving for perfection might lead one to insist that each area or country be described twice to complete the picture by the use of both methods. Not only would this be tedious but it is not always necessary for an accurate picture. As will be shown later, functional regions are not as frequent or complete as formal regions and *apart from at their edges*, description of them necessarily includes description of their formal components. Incidentally, while formal description does not necessitate any mention of wider functions, the fact that functional description does involve recognition of formal regions puts it on a higher plane and leads to the ranks which will be considered later.

There is an increasing tendency to try to give a complete picture by covering the same ground from each point of view in recent works. In addition to the well-tried sections on general physical geography followed by the regions of the country in question we now have third sections such as Dobby's[4] chapters on the social landscapes of South East Asia; Spate's[5] section 3 of *India and Pakistan* which considers the economy as a whole in contrast to the regional aspects of section 4. This balancing of general and regional geography could be developed further to a balancing of formal and functional regions in a book to give the best single description of the country concerned using the more appropriate method in any one part of it.

While men's activities on the earth's surface do not inter-relate sufficiently to make it practicable to conceive of an all-embracing functioning unit, the ramifications of some functional regions are far-reaching. So in addition to the characteristics already mentioned, unity in diversity and the dependence of one part on the others, we find that functional regions may consist of several separated areas; that they are much more difficult to isolate than formal regions; and that the functions by which a region is defined may not involve a complete description of that area.

The functional region puts the emphasis on Man's economic activities and in particular this will involve a study of how different areas work together to produce such end products as food, clothes, houses and luxuries. The physical environment becomes secondary, and of interest only to the extent that it explains the different types of farming to a 'determinist' geographer, or how different foods and raw materials *may* be produced to a 'possibilist' geographer.

This method of study which arrives last at the physical features from the desire to explain Man's activities is often urged but when actually used it is seldom satisfactory to the majority still schooled in traditional geography which deals thoroughly and exhaustively with all possible aspects of the physical world before daring to introduce Man.[6] Not only may interesting aspects of the structure, relief, climate and vegetation be ignored if they have no proven influence on Man's functions but also aspects of subsistence economies, demography and settlement may rightly be ignored in a study which takes the functional rather than the formal unit as its basis. The emphasis is on commercial production, manufacturing, trade, transport and communications, ports, market and industrial towns, the working population and consumption.

Not only is the functional region not necessarily coterminous with one unit area of the earth's surface (an area supplying and being served by the functional centre may be separated from the centre and linked only by trade routes), but it is also further complicated by being interwoven with other regions. The formal region is essentially separate and distinct, at least in the imagination if not in fact, but the description of a functional region which tries to take account of all the interplay of human actions has to deal with the results of millions of separate actions performed by individuals all thinking for themselves. It selects what is important in an area for the functioning of a giant machine. In studying the body one must at one stage consider the skeleton, the muscles, the circulation of the blood, the skin; at another stage one must consider all these functioning together in, say, an arm or a leg, but both studies are incomplete and are complementary to each other.

Finally, the very complexity of the idea of the functional region has led to several definitions. Diverse authorities variously define the functional region:

> A cultural area, an assemblage of such forms as have interdependence, and is functionally differentiated from other areas. Carl O. Sauer

> An area or unit in which the economic and social activities of the population are integrated round a focal and administrative centre.
> R. D. Mackenzie

> Organic regions may be defined as areas within which a higher degree of mutual dependency exists than in relationships outside that area.
> V. B. Stanberry

An organic region may be defined as an area whose people are bound together by mutual dependencies arising from common interests.
American Society of Planning Officials

[It is] comprised of a constellation of communities. Dawson and Gettys

[1] Spate, O. H. K., *India and Pakistan*, Methuen, 3rd edn. 1967

[2] James, P. E., *Latin America*, Cassell, 3rd edn. 1959

[3] See ch. 2 in Chisholm, Michael, *Rural Settlement and Land Use*, Hutchinson, 1962

[4] Dobby, E. H. G., *South East Asia*, ULP, 8th edn. 1966

[5] Spate, op. cit.

[6] Wellington, J. H., *Southern Africa*, 2 vols., CUP, 1955

4

THE RANKING OF REGIONS

The last definition leads on to the idea that formal and functional regions are not just two complementary aspects of the same piece of ground but are of different sizes and may result from studying and describing the ground in ever-increasing intensity.

It has already been shown that the position of a region is its unique characteristic; but, as will be shown, this is something which may be shared with a greater whole, of which the smallest region is a part. It was also suggested earlier that a certain shape is more characteristic of formal regions than of others. The geographer using the regional method tends to impose a framework of compact regions on his area in contrast to the true believer who is prepared to recognise the existence of long, thin sinuous tracts irrespective of the inconvenience of description. For the two extremes of shape, contrast Thornthwaite's map of climatic regions, revealed by a mass of detailed observations and the study of weather records over long periods all over the world, with Christaller's idealised diagram of the distribution of urban centres. In this latter, what amount to functional regions around the urban centres become perfectly hexagonal in shape. Some geographers might well like to see the world divided into symmetrical hexagonal regions like the cells in a beehive, but only in a few small areas in the Midwest, East Anglia and southern Germany does Christaller's theory approach reality.

The only reference to the size of regions, so far, however, has been the implication that they can be of any size according to the intensity of the study, with the suggestion that there is a tendency to keep the regions similar in size at any given level. The ideas of size, form and function can be combined to give ranks of regions which change steadily in all these respects: at each rank the region is larger and more complicated, but the formal and functional aspects are seen to alternate. The intensity of study is not the only factor in this system as the direction of study may give different results. By working upwards from the smallest possible unit in a given area one may arrive at a rank different from that achieved by working downwards from the world, continent or country as the starting place.

The earliest suggestions of ranks of regions increasing in size and complexity came from Passarge and others in Germany. The *Gegend* was taken as the smallest unit, usually of uniform relief, which was combined with others to form the *Landschaftsteil*. In turn several of these units combined to make up the larger region, the *Landschaft*, at which stage climatic differences were sufficiently marked to be taken into consideration. Larger units still were the *Landsteil*, combining at last to form the largest and most complex region, which certainly included all physical features if not Man, the *Land*. The *Land* might be a whole country, or a major division of the country such as the Alps in Switzerland. The *Land* at the top of the hierarchy was considered to be unique, as the Alps are, but further down the scale one small *Landschaft* of the Alps might have much in common with other mountain areas and is simply an Alpine type of region. Certainly if human activities are excluded, glacial trenches with moraine deposits on their floors and the marked microclimates of adret and ubac can be found as far apart as Scandinavia and New Zealand, but the Alps, Andes and Southern Alps, each in their entirety, are unique.

The German words given here are best used as proper names, and no attempt should be made to translate them and use what seems to be the equivalent in English. These are not landscapes but often simply single-topic or double-topic regions and the word *Land* in particular is too easily mistaken for the English. This problem is just as real in our own language, for while geographers are keen to keep the word region for this special concept, and are careful to use the words 'region' and 'area' in different senses, the former word is in common use outside geography and has no exact meaning there. Attempts to coin a name for 'regions' have met with little success; the word is too emotive ever to be dropped.

A similar but simpler system has been used by Penck who combined *Chores* to form the harmony of the larger *Landesgestalt*, for example in Germany the three Chores of Alps, plateau and plain form the Landesgestalt of the whole country. For our purposes the names are not important. At different times and places many similar systems have been constructed but with many different labels. Few of these labels have achieved general acceptance and in fact have often obscured the simple concepts for which they were presumably meant to be handy labels. What is important is the basic idea common to these systems and others such as the site, tract, section, province and continental subdivision system of Linton and Unstead, that at one end of the scale one may examine and describe the land in such exhaustive, minute detail that a division into thousands of tiny regions becomes necessary; but as one takes the wider view these tiny regions are then seen here to have much in common, there to contrast with their neighbours so that groupings are possible and desirable, thus forming fewer, larger regions which are necessarily more complex and about which one therefore generalises more and more.

However, it is not just a matter of larger regions including smaller regions and taking account of more topics. If this were so the only question would be at what level of intensity to study a particular area, whether to seek as many slight changes in slope, microclimate, soil profile and types of grass as possible, or boldly to divide Britain into the highlands and the lowlands and be content with work well done. The degree of intensity and the level reached often depends on the direction of study. Building up regions from tiny components tends to give the result of many small regions and a preoccupation with the trees to the detriment of the forest. Just as dangerously the dissection of a large continent tends not to go far enough and the reader will be familiar with regional divisions of continents where the author has seen fit to describe many exceptions and special places within his regions rather than take the division any further. Just as formal and functional regional methods are not mutually exclusive, neither are these methods of working up and working down, combining and dissecting, so that the counsel of perfection would insist on all these methods being used to arrive at the optimum balance.

Linton has given such a counsel of perfection and has put forward the clearest ranking of regions.[1] His argument runs that, starting with one of the two ultimate units in geography, working downwards dividing the continent into morphological regions, one ends up with major divisions, provinces and sections. Working downwards, using the same criteria, one can proceed no further. (Examples of these regions are given in Table 1 on p. 50.) Thus Fenneman's work, on which this is based, is adequate for the continent of North America, but not for an island or fragments of major divisions such as Britain. Therefore Linton attacked the problem from the other end and started working upwards, building his regions in contrast to dissecting a continent. For this process he lays emphasis on the other ultimate unit, the smallest area with uniform surface, the flat or the slope, which he named the site. As slopes vary in gradient, curve and aspect and the flats in altitude and extent there is a limited variety of possible sites which are found to be constantly repeated in an area of (say) thirty square miles. This concept is greatly strengthened by the work of others in different studies such as ecology and agronomy, in fact to such an extent that the site may be regarded as the fundamental unit in structure, morphology, microclimate (according to Geiger[2]), soil (G. R. Clarke[3]), vegetation (R. Bourne[4]) and farming (S. W. Wooldridge[5]). As these sites are so often repeated, groups of similar assemblages of sites can be taken to form Linton's 'minor regions' which he later named stows. In the same way groups of similar stows may be formed into his 'regions' later named tracts. Both the names stow and tract are taken from Unstead who recognised regions of both these ranks independently of Linton and without the use of the fundamental units or sites. The tract is also recognised as identical in size and character to the French *pays* and, finally, groups of similar tracts correspond with the sections already mentioned. Not only has Linton built upwards to construct regions smaller than the section but he has also used different criteria for defining each size of region. Thus the stow has 'unity by repetition of similar sites'; the tract has 'unity of form'; and the section 'unity of erosion history'. It is here that Linton is most vague; he is vague about the criteria and his only reference to the building process is to say that one takes groups of similar regions of one size to make one larger region.

This vagueness, however, is unimportant here. The important points are, firstly, that Linton found it necessary and possible to analyse and synthesise at the same time to arrive at one complete hierarchy. Secondly, that different criteria are used at different stages to define and delimit these regions. Thirdly, we have a complete, well-defined scale of regions which it is possible to apply to any land area, of any size between the two ultimate units 'given' by nature, the site and the continent. Linton himself emphasised that these are essentially relief, or more properly morphological, regions, but they are so closely connected with the other physical phenomena and with farming as to be the most generally useful regions for

TABLE 1*

HIERARCHY OF MORPHOLOGICAL REGIONS

UNIT	DESCRIPTION	EXAMPLE		AUTHOR
Continent	The unique unit	North America	ANALYSIS	Fenneman
Major division	Part of a continent with relief and morphology different from its neighbouring areas	1. The coastal plain 2. The Appalachians 3. Interior lowlands 4. Rocky Mountains	↓	"
Province	Part of a major division, divided on the same basis	The high plains as a province of the interior lowlands	↓	"
Section	Part of a province and a collection of tracts which are not all the same	The Black Hills as a section of the high plains; S. E. England including scarps and vales	↓ ↑	"
Tract (*pays*)	A collection of stows with repeating patterns	North Downs	↑	Unstead and Linton
Stow	A collection of flats and slopes	1. Scarp slope 2. Dip slope 3. Vale with undulating surface	↑	"
Site	A flat *or* a slope, the smallest possible unit indivisible on the basis of form	A regular slope or a terrace flat within the area of (say) a dip slope	SYNTHESIS	Linton

* cf. table on p. 139

geographers. For this reason, the names and the essential points of the ranks are presented here in a table. Linton defines the change from one stow to another thus:

> If in any area the physiographic conditions exist for the production of a particular slope form they are usually rather widespread and examples of that form are likely to be repeated fairly commonly over the area. . . . If one passes out of the area of uniformity of relief, rock and physical evolution, no matter which of the variables be changed, the change will be marked by the disappearance of some characteristic sites and the appearance of some that are new.[6]

In addition to this scale of size and complexity we see a combination of formal and functional regions, interrelated in such a way that there is a genuine hierarchy, the lower, smaller orders tributary to the higher, and each dependent on the other for part of its existence. In contrast to the German system, as functional regions are involved, this hierarchy must necessarily take account of human geography and this may in fact be the major element. The base of the hierarchy is found at a much lower level than that to which regional geography usually descends. It has been suggested that the smallest formal region which one can distinguish (and conversely the largest over which one can generalise with accuracy) is the field. This suggestion is a typical product of the Anglo-Saxon mind, conjuring up in the minds of kindred spirits the vision of a perfect meadow, or a neat crop of barley. The concept may not mean as much to the Celt in Connemara with cabbages, potatoes, a donkey and two cows in the same croft, nor to the Tuscan with his threefold arrangement of olives, vines and melons; but it will apply fairly well in North America, some parts of Europe, the fringes of Australia and in New Zealand. Confining the idea to its native soil for the time being, the next stage in the hierarchy is the functional unit of the farm. Perhaps the medieval three-field system or the genuinely subsistence peasant farm are the best examples of this functional unit but even the most specialised, commercial, modern dairy farm may be regarded as having its focus on the house, barn and dairy which control the working of the unit, channel the movement of commodities in and out, and constitute the most highly organised point of the area.

So far this may seem to be beneath the concern of regional geography, as stage three of the hierarchy, the formal farming region, usually constitutes the smallest unit to be described, even if the writer actually visits several farms while collecting his information. Certainly there are innumerable fields and farms within the agricultural region, but a few writers are taking account of certain aspects of these two features. Houston[7] considers the fields to be an important part of the landscape while Demangeon[8] has shown that the types of farms are essential features in the regional landscapes of France and must be taken into account in its regional division. While consideration of each field is impossible and of each farm unnecessary there is much to be gained from an inclusion of some examples from these lower orders in the final description. This idea of the case study where the actual fields of a real farm are described vividly and fully is in general use mainly in schools at the moment, with a very few school textbooks and many pamphlets and papers providing the necessary facts.[9] This use in schools should not condemn the idea out of hand. Much of the original work at Oxford by Herbertson and Mackinder on the world regions at the other end of the hierarchy was for the purposes of teaching and use in schools and there may be a chance now of ideas moving in the opposite direction. The fields and the farms exist; they are the building bricks of the larger regions, the higher orders, and while each brick cannot get and does not need individual attention, the architect ought to know its shape, size, composition and function.

By stage three, the area in which all the farms are sufficiently alike to make accurate generalisation possible, we have returned to a formal region for in fact it is extremely rare to find a vividly contrasting variety of farms mixed together in one small area. Each field on a farm may have a different use; but however little determinism there may be, individual farmers making their choice in an area from all the possibilities are guided by relief, climate, soil, tradition, business organisation, subsidies and the state of the market to produce much the same as their neighbours *under the same conditions*. Thus, in fact, we can distinguish formal agricultural regions where the similarities between the farms, especially in the produce, or the crop combinations, override everything else.

So far, one type of region has not cut across another; the one has been included completely in the other. Complete fields make up the farm, complete farms make up the region. But at the next stage in the hierarchy it is not necessary that all of any one formal agricultural region be included in the functional region which is one stage higher. At this level one may imagine a market town in a typical situation at the junction of two formal regions: between the Pennines and the Yorkshire plain, on the eastern edge of the Rockies, along the Alpine foreland, and so on. Such a market town exists to serve these regions, and thrives because of the extra trade and wealth provided by a contrast of products, making exchange necessary and desirable. But one market town may not be able to serve *all* the area covered by the two regions. The towns may vary in size from Richmond to Denver but so do the regions and only part of each of the regions, mountain and plain, are involved in the functional region of the town's service area. Thus many towns are found in lines along the junctions of these contrasting formal regions and the functional zones overlap them and may be imagined to cross them at right-angles (Fig. 2). On examination it seems that very rarely are two formal regions of this nature coincident with one functional region of the higher order. In fact on the other side of the Pennines and Alps there exists a similar arrangement so that the western Pennines are tributary to Lancashire, the eastern to York-

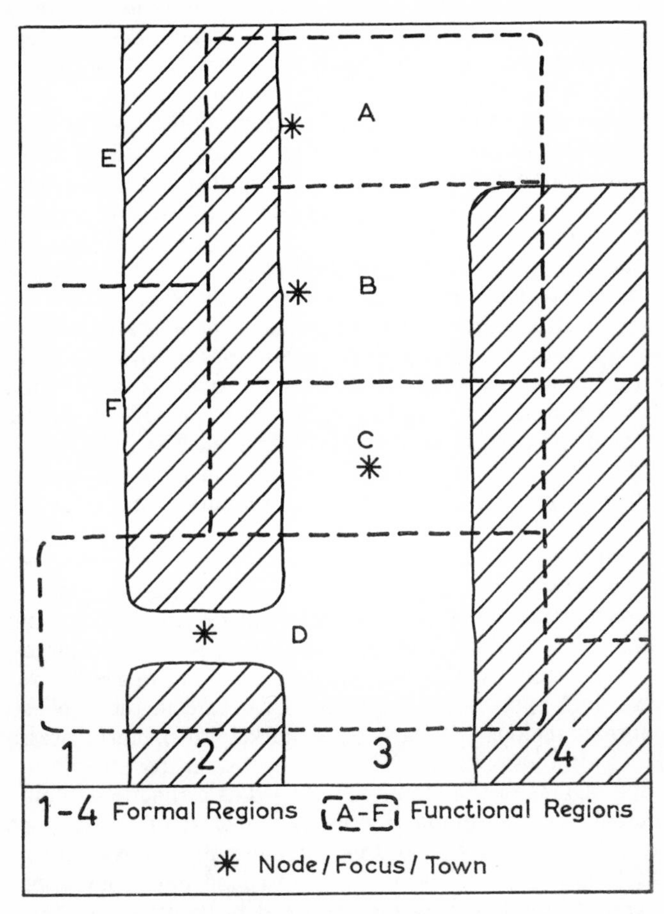

Fig. 2 Functional regions. Each combines parts of the formal regions represented by the bands 1–4

shire, parts of the Alps to Switzerland, parts to the Po valley. So that the mountain region in each case can be regarded as a homogeneous, uniform formal region while in fact one end of it may direct its trade westwards and be involved with the lowlands to the west, while the other end is separately involved eastwards. The two parts of the same formal region are thus often found to have much more in common with parts of other regions by reason of their functioning together than they have with each other. This unifying action of a town and the ties of trade and self-interest are often much stronger bonds in real life than mere similarities in landscape and daily life which may be apparent on the map or on paper but are not regarded as important on the ground.

The next stage in the hierarchy has been implied already. If there are several towns at the junction of plain and upland, each with a functional region, these can be grouped together into a formal region of a higher order still. By now we are dealing with large portions of any country and are reaching the top of the pyramid. Having started with the tiny, formal region of the field, with alternate functional and formal regions, the final region which embraces the whole country must be considered functional. Such countries as Denmark and Egypt come to mind where the land use is so nearly uniform that it seems almost possible to regard the country as a formal unit, but in the majority of cases the regions of a country are sufficiently diverse. In some cases, as in Britain and France, it is then possible to find each of the formal regions of the highest order combining into the one functional region which is tributary to the metropolis.

In countries such as Africa and Australia, however, this certainly is not possible for in neither case can one city be considered the metropolis. Even in Australia which is united politically there are in effect several functional regions each focused separately on one of the half-dozen major cities, and Sydney is simply the largest of these, not a metropolis in the fullest sense of the word. Where this concept does apply there is still need for care in the application. The formal regions tributary to a market town must necessarily contact it. But just as some fields may be distant from the farmyard, so some regions of that country regarded as a functional unit may not actually contact the metropolis. The picture of regions fanning out from a central city or arranged in concentric rings is much too simple. The regions of the Paris Basin round Paris may approach this ideal but the Pyrenees, the Midi and Brittany, though separate and distant, are as important parts of the whole as any others.

Whyte's investigation of the interconnections between houses in suburbia[10] has suggested that a similar hierarchy may be found in towns. We may recognise one room in the house, or one shop in the factory as the smallest possible formal region. Then any house, factory or public building will be considered as the smallest functional region, with its various parts organised for some purpose of living, production or service. Groups of houses, factories and some public buildings will in turn make up the next formal regions in the hierarchy. Residential and industrial zones are readily recognised in any town or city, but at this point the interests of Whyte and some of the extreme social geographers inclining to sociology proper diverge from those of the regional geographer trying to balance all the elements. Whyte distinguishes the street and the neighbourhood which seem unnecessarily fine distinctions for our purposes. Certain neighbourhoods may have unique characters in appearance, interests and social standing in the residential regions but these social groupings are outside the scope of regional geography. If sufficient public buildings are grouped together a business region may be distinguished (those of banks, insurance, merchants, lawyers) or a cultural region (the larger schools, theatres, galleries, museums and possibly hospitals) or, in a different arrangement, an entertainment or recreational region (theatres, cinemas, clubs, bowling alleys, restaurants, parks): the only essential is to find enough buildings with similar functions to make a formal region anywhere in the town. In the small market town usually it is possible to recognise only a simple concentric arrangement of the town centre with market, shops and services, the inner ring of high density poorer dwellings and the outer ring of larger houses in the suburbs spaced further apart. In the large city, however, and certainly in the conurbations and the metropolis, there exist towns within towns and the idea of the neighbourhood as a technical term may have some significance here. Many parts of London have their own little 'town centres' consisting of shops and buildings such as are found in the genuine isolated country market town. In the case of London they are surrounded by a maze of overcrowded residential streets but the zoning and even the conscious community spirit are there.

To see how complex this weave of formal warp and functional weft is in the final fabric of the hierarchy of regions we must examine this urban division from both ends. Working upward through formal room, functional building, formal residential area to functional neighbourhood or town, as in the case of farming regions we arrive at a functional region at the top of the hierarchy. Working down, however, breaking the country into its component regions, having arrived at the city with its surrounding formal regions making up a functional unit, isolating the city as a type of landscape from the surrounding market gardening region and dairy farming region, it contrasts with these as a *formal* unit. The city is both formal and functional at one and the same time according to how the geographer regards it. At one level of study the city presents simply a formal urban landscape contrasting with adjacent arable or pastoral landscapes. At another level the city is a complex organisation of contrasting complementary parts each vital to its working. From this it would seem that at least regions of this type are equally products of the geographer's mind and of events in the world outside.

To complete this consideration of functional regions, two points must be added which lead to the conclusion that where such regions exist they are incomplete. If, as has been shown earlier, functional regions are fragmented, then it follows that there are areas between the parts of the region which have no connection with it. In fact or in imagination these *may* be themselves parts of another functional region occupying the same ground, interleaved with the first but separate from it. But it is possible that the human activities in these areas, at least those usually selected by the geographer, have no significant connection with other areas elsewhere. Whyte[11] mentions examples in the suburbs which he studied in order to find functioning neighbourhood groups. The people of one community who share the same interests, church, clubs (and most important for Whyte the same kind of work in a large organisation), while living close together in a certain type of neighbourhood, obviously do not all live in adjacent houses. The invisible social ties stretch from one house to another forming the functional region. Yet Whyte found that people of the same or similar organisations, with the same interests and way of life, were completely isolated in the middle of this community, cut off by such chances as houses facing the wrong way, or being off the main road. This adds to McCarty's evidence that areas in a functional region are not necessarily part of it; they are like interstitial cells in some organism.[12] Another suggestion on similar lines comes from Smailes[13] that hamlets between very widely spaced towns are in a 'service vacuum' (Fig. 3 and Appendix).

Secondly, functional regions change much more rapidly than formal regions where the physical components give greater stability, and a fundamental contrast between the two types of region is that a functional region can grow and shrink in areal extent while a formal region is much more static. Ignoring exceedingly long-term morphological changes, and doubtful climatic changes which have yet to be proved, it is true that (say) a formal agricultural region may extend and retract as did the wheat-farming region on the High Plains in the 1920s and 30s. But this change in the area under one crop, or one type of farming, will result in its coincidence with or

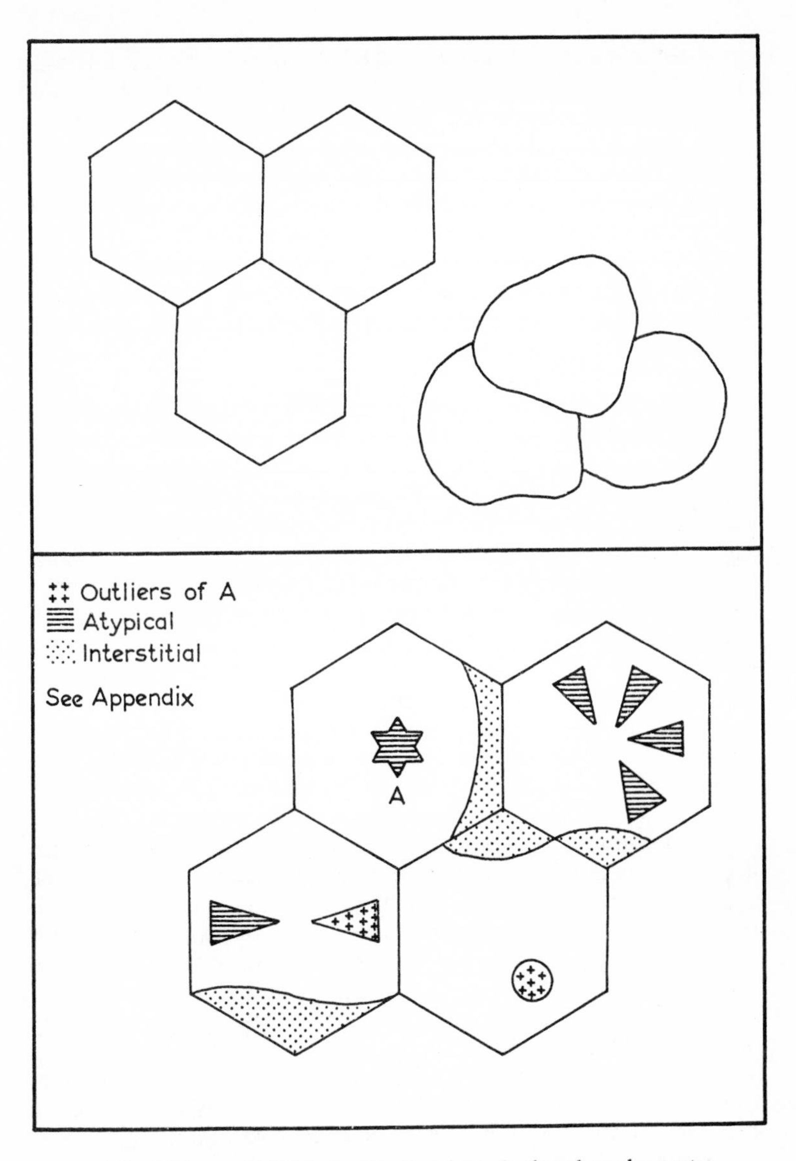

Fig. 3 The areal discontinuity of regions. Rather than the neat pattern of the upper diagram, in reality one often finds some arrangement as untidy as in the lower diagram

cutting across some other one-topic region, say of rock, relief or soil, so that the same regions can still be recognised. In the case of the functional region, however, the trade and ramifications of some small industry may grow from one town to involve widely scattered parts of the whole world. For example, the car industry has grown from a few isolated garden sheds in Europe and the USA to some of the largest organisations drawing iron ore, non-ferrous metals, textiles, rubber, etc., from all over the globe and exporting to every country. Concurrently the British cotton textile trade has shrunk from its imperial size until it can be further harassed by tiny Hong Kong.

Moreover, the functional region changes in character as well as extent. Considering the whole hierarchy it is possible to see the larger regions near the top of the pyramid rearranging themselves on a larger scale. Since the agrarian and industrial revolutions regions have been changing from subsistence to specialist commercial economies; from now on, however, the trend would seem to be a return to national subsistence. First was the stage of local subsistence in the feudal village or the ladang clearing depending on a variety of products in a small area; secondly the specialisation in dairy farming or rubber production which depended on world trade and variety of production within the largest area. Thirdly, now that countries which traditionally supplied food are developing industries and feeding their own people instead of the world at large, heavily industrialised countries are concerned for their food supply and there is a trend to variety of production within a smaller area again, within the national boundary and customs barriers. The Common Market is an example *for* this argument; as tariffs are lowered between the Six, they are raised against the rest of the world, making little Europe an effective economic unit of a size to compare with the USSR and the USA. Thus one area of constant size during this period may have been first a self-contained subsistence region; secondly part of a world-wide functional region; and now be reorganising for subsistence again. Borrowing from an early, very useful concept in geomorphology, that of 'process, structure, and stage' put forward by W. M. Davis, we might recognise the functional region as one stage of a process acting on the resources of the physical world. In this case the process is the sum total of Man's economic activities, and like the climatic processes of erosion can change in time as Man's aims, values and abilities change. Hence the functional region is a rapidly changing phenomenon.

[1] Linton, D. L., 'The delimitation of morphological regions', ch. 9 in *London Essays in Geography*, eds. L. D. Stamp and S. W. Wooldridge, Longmans, 1951

[2] op. cit.

[3] Clarke, G. R., *The Study of the Soil in The Field*, Oxford, 1938

[4] Bourne, R., *Regional Survey*. Oxford Forestry Memoirs, no. 13

[5] Wooldridge, S. W., *The Land of Britain*, no. 51, *Yorkshire* (*N. Riding*), London Geographical Publications Ltd, 1945

[6] Linton, op. cit.

[7] Houston, J. M., *A Social Geography of Europe*, Duckworth, ch. 4

[8] *Géographie Universelle*, tome VI, *Géographie Economique et Sociale de la France*, vol. 2, ch. 7

[9] See Cumberland, K. B., *South-west Pacific*, Methuen, 2nd edn., 1958

[10] Whyte, W. H., *The Organisation Man*, Cape, 1957, ch. 25

[11] Whyte, op. cit.

[12] McCarty, H. H., *The Geographic Basis of American Economic Life*, Harper, 1940

[13] Smailes, A. E., *Geography*, vol. 32, 1947 and *T.I.B.G.*, vol 11, 1946

5

REGIONALISM

The theme of the first two chapters was that there is still doubt as to whether regions exist or are simply useful devices, but there seems little doubt that the city-region as commonly described is a product of the imagination with even less visible basis than the formal region. The concept of the city-region has been put forward as much for the purpose of creating regions as for simply describing regions. In this book the city-region will be classed with all other regional work which has some political motive as 'regionalism'.

A nice distinction must be made between the city-region and the urban field. The town and the country form one unit, one variety of functional region, which in Britain is called the urban field and which is essentially an economic relationship between the two. In contrast the city-region exists (some would say *should* exist) to serve the social needs of the people. Dickinson[1] writes of the city-region as a natural social unit, or community space-grouping, where the emphasis is on the social unit. His book seems to have been misunderstood by some geographers as it is often referred to as though it were a work on cities, regions and regional geography in general. Yet Dickinson confines the book to city-regions and it is in the International Library of Sociology and Social Reconstruction where it quite correctly belongs for it is concerned with sociology and with planning for, and the creation of, one type of region. Christaller, Smailes and even Dickinson earlier, mapped urban fields as they found them, but in this book Dickinson, like the planners, hoped to make city-regions what he wanted them to be. As a policy, this may have much to commend it, but things do go wrong when objective description and purposeful propaganda can easily be confused as one and the same thing.

Dickinson's main interest is in providing for the full social life and one of his most important themes is that there should be plans to provide the social amenities which rapid economic development has neglected. So between the recorded facts of Smailes and the dreams of Dickinson we have the half-world of city-regions and regionalism. Others have seized on city-regions as a much better concept than formal regions and have tried to divide various countries into such units. For example E. G. R. Taylor[2] attempted to divide Britain into city-regions and was only partially successful because when all possible regions have been marked there are large gaps. Moreover, these regions were defined by drawing circles as big as seemed reasonably sensible round each large city. There is enough in this to make the regions plausible; towns such as Plymouth and Norwich obviously do provide for most of the social needs of the surrounding rural population, but it might be hard to prove to people in Market Drayton that Manchester is their cultural centre. Obviously, drawing city-regions does not make city-regions.

Some of the counties and county towns of the English Midlands have been city-regions for centuries, but industrial cities such as Leeds and Bradford have not really been part of such functional regions, while the highlands and Wales have never had a central city which could serve them in this way. Thus it is wishful thinking to draw a city-region where a large city is surrounded by a convenient rural area unless the city and countryside are proved to have such social relationships as Dickinson describes. Moreover, his criteria are not those normally applied by a regional geographer. The centres of his regions are where social ties and social activities are strongest, the borders are where these are weakest, but how does one measure entertainment, recreation and, above all, meeting people and visiting their homes?

Regionalism has been defined above as the concept of the region used for purposes other than geographic description, so that the motives behind them must be given some consideration. Some politicians, businessmen, architects and town planners have seized on the idea of the functional region, and especially the city-region, to further their own ends. Regionalism uses the concept as a means to such ends as private gain or public power, where geography uses it as a tool for research and a vehicle for description. Some regionalism has been praised by geographers, particularly the French movement to emphasise characteristics of the Pays and to combat the domination of Paris. In contrast the emphasis on regional characteristics is not universally praised in the USA. While the Federal Government is trying to hold the country together by stressing the common culture and purpose of the nation certain States openly defy it. They stress their local problems and customs, especially in economic and racial matters, and this tendency to fragmentation is called sectionalism, which implies self-interest and the use of regional arguments to hide the true, less justifiable motives for a greater degree of independence. Similarly the declining industrial regions in Britain stress their regional unity most to try to stop the decline, and we hear much more about the regions in the North than about the Midlands and the south-east where booming industry is attracting more and more people.

Geographers may well be unwittingly responsible for this regionalism because they never use their regions for any other purpose than description. This is where the practical politicians and planners step in with their attitude that if regions exist they must exist *for* something and therefore must be used. Professor Gilbert[3] hinted at this when he said that such men are out 'to develop in people some regional loyalty', but he did not pursue the motives for this. Where geographers have produced a convenient regional division this may be put to practical use for administration or planning. Gilbert gives examples of such divisions from several countries and, in addition, some made for specific government or planning purposes. Those geographers who share Dickinson's ideals might consider that one function of geography may be to find and describe latent city-regions ripe for development, and to provide perfect examples which do exist as models for this development. For, as Gilbert stated, 'Geography is the art of recognising and describing the personalities of regions.'

This better type of regionalism may degenerate into sectionalism, the fragmentation of a country and the emphasis of the town at the expense of rural areas. There are three common arguments in favour of this. Firstly, that if the metropolis is the only control centre in wartime its destruction would paralyse the country at once. Therefore city-regions are desirable, to be self-sufficient in wartime. Secondly, that as governmental functions multiply, more work should be delegated to local authorities who can adapt central policy to regional conditions. Thirdly, that the old-established boundaries of counties and towns should be readjusted to fit modern conditions. Gilbert criticised the arrangement whereby the town centre and its suburbs are often administered by separate bodies such as the borough council and parish council. Each authority deals with its own problems better but the work is doubled and there is often conflicting interest. A single council might govern the whole region more efficiently, but perhaps not by resolving the interests; rather by ignoring the needs of the suburban areas with fewer votes and lower rates. The drawbacks of both systems are illustrated by the facts that while Lancashire and Cheshire County Councils are considering joining together to form one functional region centred on Manchester, the Sheffield authorities are finding that there is too much centralisation in Yorkshire, that Sheffield's interests are being neglected and that the city should have more control over its own area in the south. It is important to realise the motives behind this regionalism, to sort the worthy from the hypocritical, and to keep regionalism separate from academic geography.

Professor Hartshorne[4] pointed out that some geographers are determined to define regions, whether they exist or not, in order to turn geography into a pure science complete with objects of study (regions) which will offer the possibility of generic concepts. This is clearly regionalism, the use of regions for some ulterior motive, because to such geographers the fact that regions are objects of study, or vehicles for description, is secondary to the fact that if they prove regions exist they have raised the status of geography and increased their own importance. This scientism,[5] or desire for scientific status, led to the suggestion that regions are organisms by analogy with Biology and from the use of organic metaphors.[6] In fact the attempt was made to see the earth's surface as a mosaic of functional regions which are organisms. Hartshorne's clinching argument is that regions cannot reproduce themselves, therefore they are not organisms, and that there is not even any evolutionary connection between them.[7] While Hartshorne dismisses the idea of organisms completely, the *comparison* of regions with organisms can be of some help. By analogy with the human body we may make clearer the interdependence of some regions without implying that every part of the earth's surface is part of a living organism. Similarly the knowledge that two regions have been colonised by the same type of people may be the key to understanding them. An observation such as Dobby's[8] that much of Java's farming is in fact the tropical version of Dutch market gardening at once makes his description more meaningful and emphasises the combination of Dutch and Indonesian ancestry in the evolution of the present landscape. Any method which helps comprehension without distortion, and is succinct and vivid, is too useful to condemn out of hand.

There is at least one respectable motive for regionalism, for one outcome of an enthusiasm for city-regions may be a much better use of the land. An original belief in a city-region may result in the achievement of such a thing in real life. This reverses the geographical method of observing and then describing the earth's surface. In this case the idealised description may come first, and the actual landscape be changed to fit the ideal. Perhaps the most comprehensive schemes put forward for ideal regions are those of Lewis Mumford[9] but, while these have not been put into effect, other more limited schemes have been tried. Apart from the Communist-inspired industrial and farming regions of the USSR[10] it seems that a completely new start has been necessary for these schemes which are characteristic of the New World, in its widest sense, rather than the old. Some examples are the Mormon State of Deseret in modern Utah; Christchurch in New Zealand, the product of the Canterbury Association; the Orange Free State and Liberia. Obviously it is harder to bring a planned region into existence where a cultural landscape already exists, and movement to a new land sorts out the people as well as giving virgin territory on which to build. Thus the dreams of turning British and European cities and their surroundings into city-regions will be very difficult to realise.

However, at least two dreams are becoming reality in most unpromising areas and new regions are being created which geographers will have to describe. The polders created from the bed of the Yssel Meer are now facts, with farms, roads and model villages on them. The *kibbutz* of Israel now impose Man's regular patterns on the desert, and one type of co-operative community, the Moshav Ovdim,[11] is the living, tangible microcosm of the city-region. Such villages of about a thousand people are sometimes circular, a good arrangement for tackling the deserts, similar in plan to medieval Terp villages adapted to forest reclamation. The inner ring of the village contains the communal buildings to serve the co-operative economy and the social functions. The outer ring consists of the homes of independent smallholders and radiating from each farmhouse is an ever-widening segment of cultivated land which ends abruptly against the desert (Fig. 4). There is a twofold interest here, in that we have a model of the much more complex city-region and in the fact that such an isolated functional region has been thought out and built by people who are conscious of it as a whole. Moreover, it exists as an undeniable object of which geographers must take account.

In most parts of the world the geographer has been the first to

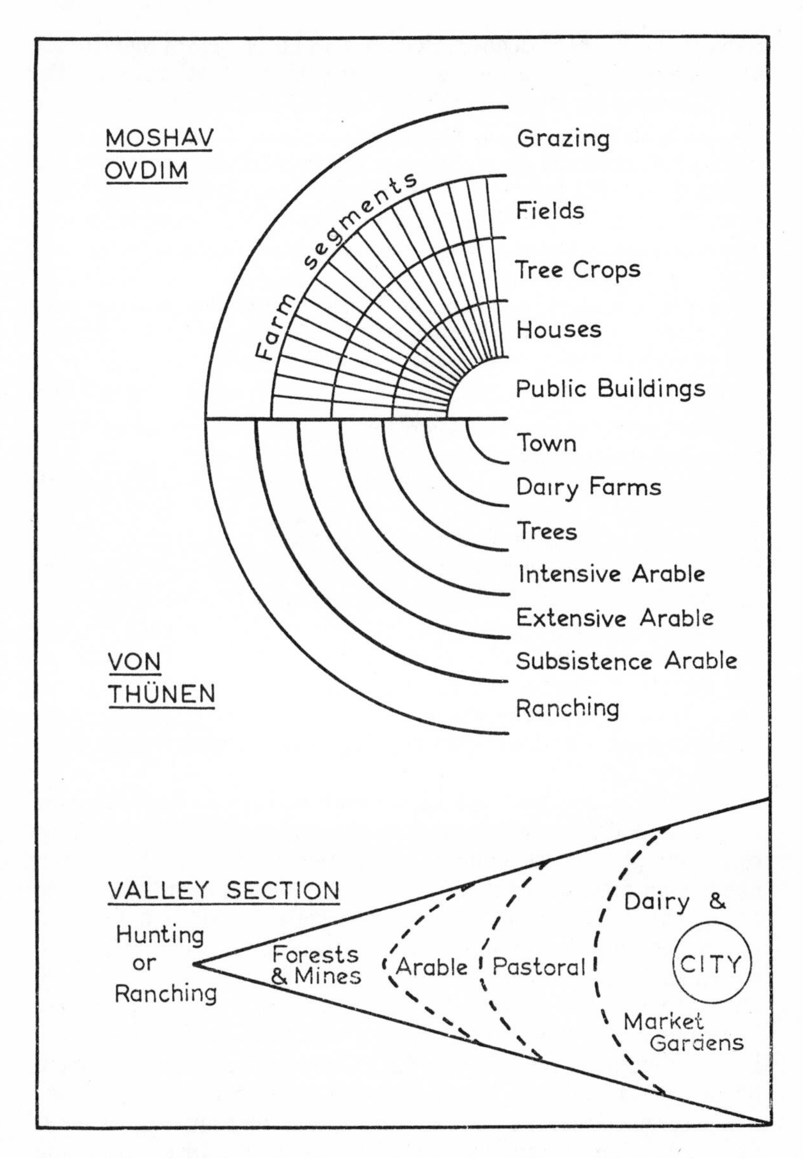

Fig. 4 Diagrams to illustrate the Moshav Ovdim, Von Thünen's theory and Le Play's Valley Section

name and attempt to define regions but, as in the case of the Moshav Ovdim, there are some areas where the regional consciousness of the inhabitants came first, especially where they set out to establish a special way of life, and the geographer becomes aware of this later. There is some suggestion that the interaction of geography and reality should be taken further, to the extent that geographers should strive to awaken regional consciousness in people. Some definitions of regions imply this regional consciousness; a region is:

> Any one part of a national domain sufficiently unified physiographically and socially to have a true consciousness of its own customs and ideals, and to possess a sense of distinction from other parts of the country.
>
> J. Royce

> A familiar place, where we know the lay of the land, the traits of the people and their resources, needs and problems.
>
> The American Society of Planning Officials

Moreover, Professor Gilbert[12] implied that geographers should stimulate regional consciousness where it is lacking when he stated: 'the novelists have made the people understand regionalism and the region more effectively than the geographers'. This is true, but it is not usually the geographer's aim to make the people understand regionalism. If geography incidentally makes people aware of better possibilities so that they strive to make better use of their regions, well and good; but this should be a function of regional geography, not its purpose.

Perhaps the only real regions are fields and countries which have definite boundaries and of which everyone is aware. Not everything, especially physical phenomena, conforms with these boundaries but human activities fit them more closely than they fit other types of regions defined more vaguely. The better grasp man has over nature the more he organises the landscape with fences and boundaries and this may lead to the proposals for the establishment of official regions and provinces for the more efficient use of resources. In Britain the proposals have ranged from the seventeen regions of E. W. Gilbert, sixteen of E. G. R. Taylor, twelve of C. B. Fawcett, ten of G. D. H. Cole through dozens of governmental divisions down to the seven of the GPO. The number of different divisions suggests that those of the geographer are of little use to the government, and vice versa. Similarly Dickinson's concept of the city-region may be of little use to the geographer because by definition all city-regions are similar, and are best grouped together as making up one formal region of many smaller, similar, functional regions, if they are found to exist at all.

[1] Dickinson, R. E., *City Region and Regionalism*, Kegan Paul, 1947

[2] See Taylor, G., *Urban Geography*, Methuen, fig. 179, p. 378

[3] Gilbert, E. W., ch. XV in *Geography in the 20th Century*, ed. Griffith Taylor, Methuen, 1951

[4] Hartshorne, R., 'The Nature of Geography', *Annals of the Association of American Geographers*, vol. 29, pts. 3 and 4, Lancaster, Pennsylvania 1939, ch. 9, p. 254

[5] Whyte, W. H., *The Organisation Man*, pt. one, sect. 3

[6] Hartshorne, op. cit.

[7] Hartshorne, op. cit., p. 260

[8] Dobby, E. H. G., *South-east Asia*, ULP, 8th edn., 1966

[9] Mumford, Lewis, *The Culture of Cities*, Secker and Warburg, 1938; see also *Technics and Civilisation*, 1934 and *The Condition of Man*, 1944

[10] Szava-Kovats, E., in *Soviet Geography; Review and Translation*, American Geographical Society, N.Y., vol. VII, no. 7, September 1966

[11] See *National Geographic Magazine*, vol. 127, no. 3, March 1965, pp. 418–19, and Gaury, G. de, *The New State of Israel*, Verschoyle, p. 208

[12] Gilbert, op. cit.

6

THE NATURE OF REGIONAL GEOGRAPHY

The regional method has been successful so far, and has been universally adopted, but now that it seems incapable of much further progress or development, its inherent shortcomings are being emphasised and certain critics have said that it should be rejected completely. Some criticisms are valid as there is no question that the regional method has severe limitations, but its achievements are so valuable that they deserve a better fate than fashionable contempt. The real danger lies in accepting all the criticism too readily and trying to make the method more precise than in fact it can ever be. Rather than try to force regional geography to carry out both functions of giving a description of main characteristics and a full precise account of all the phenomena (which is the function of systematic geography) it will be more useful in the long run to admit the shortcomings but also to state quite clearly what the regional method can and should do. There is a danger of falling between two stools. The regional method will not work equally well in all parts of the world for all the topics usually considered. It will attract even more contempt if geographers try to define the indefinite boundaries, make regions and city-regions where none exist, and give an exact, pedantic statistical account of all the phenomena in every square mile of a continent. But this does not mean that the regional method should never be used anywhere. Once its limitations are admitted and stated, it can be used as one of a number of methods where it is more appropriate.

Regional geography lays itself open to criticism the more it tries to copy the natural sciences and the less it has in common with history and the humanities. The main works which examine the nature of geography as a whole[1, 2] consider only the purpose of the professional geographer, and in particular what aspects of the earth's surface interest him. There is never any mention that at least one purpose of geography should be to pass this on to the general public. This attitude is typical of the mystique surrounding the natural sciences and in fact is one essential difference between the humanistic academic outlook and the material, essentially practical one. Those who research into the facts of the physical world or of human society share the prime motive of the need to satisfy their own curiosity. But there the motives and aims diverge. The scientist seems more concerned then to pass on his newfound knowledge and newly evolved theories to other colleagues, for their interest, for their help, for their use, and one is left with the impression that science is only for scientists.

In contrast many historians, in particular, seem equally concerned to share their knowledge and opinions with the rest of educated mankind through their books. Taking the most cynical view, this may be because nowadays the historian is not paid for pure research, as are some scientists, and therefore he must publish in order to live, but this concern to pass on the knowledge, and the ability to make the story vivid and readable, result from a very long and respected literary tradition in this field. There are no people trying to make use of this knowledge, and no middle-man is needed to rewrite history in order that it may be understood by ordinary people with the normal education. In fact to elaborate on the points that politicians could well pay much closer attention to history, and that history has inestimable value in education, would be so much wasted time and space. In short, most people share an interest in history with the professional historians and there is no impression that historians are a race apart.

The impression that geography should be written only for other geographers becomes very strong, not only as one reads such critics as Hartshorne, but even more as one takes an objective look at the amount and type of new regional geography. Compared with any recognised discipline in schools and universities, and especially with history and natural sciences, geography is obviously not as popular, either with writers or the general reading public. The numerical evidence is there in the publishers' catalogues and in the bookshops where one may compare whole walls of shelving given over to history with the odd shelf in the corner where geography is giving way to geology. Several references have been made to both the genuine attempts to make geography more exact in its measurements and to the attempt to make it appear more 'scientific'. Whatever the motives of the writers, the effect is the same on the readers, because it puts them off. Thus this type of regional geography may fail by falling between two stools. It can never be an exact natural science because of the nature of its subject matter, but the attempts to achieve this status appear to result in its losing all its value and appeal as a humanistic study akin to history. Few objective critics can be impressed by the façade which has been erected to present this 'scientific' appearance. There are far too many maps, diagrams and tables which are completely unnecessary to the text. One may readily find in current regional geography texts the two extremes of maps which show everything, and make nothing clearer, and maps which are so simple and obvious as to be laughable, if geographers did not take them so seriously. The maps tend to be cluttered, the diagrams irrelevant and the tables singularly uninformative, but they are put in the books like buttresses round a shaky building when in fact they are more likely to pull an otherwise sound building to the ground.

Equally significant is the lack of photographs, which need to be in great profusion in even the most academic regional geographies today, when people are conditioned to the illustrated Sunday papers and the television. If photographs are specially taken, particularly from the air, which gives the geographer's eye-view of the maplike landscape, they can give a genuine extra dimension to the description. But photographs are costly to print; geographers are only just now learning how to draw maps and employ cartographers, let alone take perfect pictures or work with photographers; and, worst of all, illustrated geography books look like travel books. It is quite clear that there is little attempt to interest the general public. There is little anxiety to win over a wider public for geography, very little desire to share the excitement and interest on the part of the geographer. Then for whom are the books written? Surely not for the other professional geographers. Surely they are not taken in by the humbug of scientism and really believe that all these maps, diagrams and statistics which terrify the layman are vital and essential. If not, we are left with the in-betweens to form this small market, students, politicians and businessmen. For over a century political and commercial geographies have recommended themselves, on their dust jackets, as containing essential knowledge for the statesman, empire builder and entrepreneur. But it is fairly clear that fewer such books were sold to these shrewd men than to the traditionally poor students. The attempts to make contemporary geography books look more like physics and less like history may be perfectly in tune with the fashion of the times but there is no evidence that science students buy more books than history students, or more particularly, that geography students in a university where they belong to the Faculty of Science buy more books than those in a university where they belong to the Faculty of Arts. Anticipating the reply, if the object of the exercise is not to sell books, then geographers should remain mute, inglorious, unpublished Miltons.

This question of popularity and sales may be a minor problem in comparison with the problem of growing criticism of, and dissatisfaction with, regional geography, but both problems can be cured by the same remedy. That is to redefine and reaffirm the descriptive, subjective, personal nature of regional geography. If regional description were the whole of geography, this would be retrogression indeed but, as regional geography is limited to what it can do satisfactorily, at the same time systematic general geography can be developed more in those fields where accurate impersonal measurement and recording are being regarded as increasingly important. The complementary roles of these two branches will be considered later, but the point to be stressed here is that the re-establishing of regional geography in the literary rather than the mathematical field will at one and the same time remove it from the valid criticism that it is attempting what it is not able to do; and at least give it the chance to become as popular and as respected as history with those members of the general public who bother to buy or borrow books and read outside their particular trades. There is only a slim chance of this, because in comparison with history, geography lacks the enthusiasts who are just as anxious to interest their readers as they are to find out. It is no use wishing for a Macaulay, Trevelyan, Bryant or Churchill, and the argument must be not that such men could breathe life into regional geography, but that if regional geography confines itself to those aspects of the subject which it can cover perfectly and where it has done so well up to now, it may attract men of similar intellect and artistry.

There is a much more fundamental point which separates geography from history just as completely as it is separated from natural science. Places are interesting, but people are fascinating. A geographer might not agree with the latter statement and many scientists would agree with neither, but the evidence that most people find this so is overwhelming. By its very nature history has some story; there are dramatic movements of groups of people, and the condensed and therefore exaggerated lives of the most important men and women. But there is not necessarily any story or progression in geography. In some cases historical developments may be an essential part of the description, and determinist geography which shows how certain relief, climate and soils lead to certain farming and ways of life is still being written. But usually the description of a region is a static picture, lacking the drama and conflict of history and the fictional film, relying on the intrinsic interest of the subject matter which Grierson, Flaherty and Rotha found so satisfying in the documentary film and which the geographer finds in his subject.

It seems that the Hollywood hero will always be more popular than the Man of Arran; the Wars of the Roses more popular than the geography of Lancashire and Yorkshire, for the fictional character and the historical figure are more interesting to the majority of people than are other working people just like themselves, whether in another job, another country or another century. The documentary film, social history and regional geography are of minor importance simply because of this common subject matter of large groups of ordinary people who have to earn their living and who are important and worth study only because of their sheer numbers, and the fact that together their work produces all the wealth and substance of any country or any century.

In addition to concentrating on masses of ordinary people at work, rather than on a few rare individuals engaged in some enterprise of honourable, dangerous consequence, regional geography requires a much greater intellectual effort by the reader than does history. Thus to the layman it is not only dull, it is hard going as well. The reason is that unlike a history, the regional description has all its rewards at the end. The general reader may read any well-written history simply as a story, simply for the interest of the events as they happen. The mental process can be that of picking up, being aware for the moment, and then setting down; the eye moving over the page is like the stylus moving over the gramophone record—what is past can be forgotten. This is putting reading at its lowest level. Reading history this way is like reading a paperback thriller, whether Sherlock Holmes or Perry Mason, simply for transient amusement. But this is virtually impossible with regional geography as it is usually written. To the layman reading through the inevitable early chapters on rock, landforms, climate and soil there is no immediate reward of intrinsic interest; not in the sense of the interest to be had from a book on geology, geomorphology or general climatology. The only reward for this present work is the future synthesis right at the end of the book. The mental process must be one of continuously taking up and carrying an increasingly heavy and more complex load until all the components are available in the mind for the final synthesis, the mental concept of that particular region. As one reads *The Conquest of Mexico* one may enjoy the account of Cortéz marching through the night and fording the river to take the rival Spanish newcomers by surprise, and later forget even that vivid incident without any detriment to one's enjoyment of the rest of the story, as a story, or without any serious distortion of one's final comprehension of the conquest as a whole. The prosaic accounts of relief, climate and soil, however, are bound to make less impression, and if each chapter is forgotten while one is reading the next, then the later accounts of farming, mining and manufacturing will be less easy to understand and the final comprehension will be seriously impaired. All this is not to imply that serious students of history do not and cannot build up a synoptic view; they can and do, but history, like a Shakespeare play, can be enjoyed at many levels. The point is that anyone who bothers to read the present type of regional geography *must* make the effort to build up a synoptic view to the end if he is to get anything out of his reading except, perhaps, a few unco-ordinated scraps of general knowledge. Therefore any efforts to change the nature of regional geography would be better directed to making it genuinely more readable and comprehensible than to making it appear more 'scientific', for its own sake as well as for its popularity. However, it must be more readable in the sense of a more logical arrangement of the information and not in the sense of over-dramatising real life for the sake of vivid description at the expense of accuracy.

With the growing criticism of the capabilities and accuracy of the regional method, systematic geography is becoming the more important and respected branch of the subject. The reasons are obvious enough in a period when it is fashionable to be objective, technical and practical about everything, including subjects which are not best dealt with in this way. Systematic general geography is more exact, in that an economic geography, for example, can divide the world into single-topic regions based on the hard facts of agricultural or industrial statistics; it can be more technical because a systematic geomorphology, climatology or economic geography will explain the principles of the subject and this would lead to repetition in regional geography; systematic geography may also be useful, for once one has learned these principles they may be applied elsewhere; knowledge of physical geography is useful to mining and civil engineers, economic geography to businessmen, and social geography to planners; but above all systematic geography has the virtues of being logical, exhaustive and easy to comprehend. Those, in fact, may be seen as its main aims. A systematic account of one topic throughout the world aims to disentangle this topic from all irrelevant topics and to set out the facts in a logical way. The better works also aim to be exhaustive at least to the extent of considering all significant variations of the topic if not every example of every variation, and one expects to be able to use the better, modern systematic geographies as reference books. This latter point connects up with their greater ease of comprehension compared with regional works. A systematic geography is easier to understand for several reasons. First, it deals with a limited range of phenomena. Second, the cause and effect of physical phenomena are easier to discover and understand than those of human activities while, especially in economic geography, complex human motives are ignored. Thirdly, the synthesis is not left to the end, in fact such books often end up with more examples; one may pick a chapter on coastal erosion, tropical climates, the textile industry or the sites of towns without having to follow the arguments in the other chapters. Fourthly, there is no question that this type of geography gives an artificial view, perhaps artificial in its simplicity but certainly artificial in the isolation of one phenomenon from complex reality.

Because these aims are regarded as more worthwhile at the moment, some regional geographies seem to be making these aims their own, and so losing all justification as a branch of the subject in their own right. As stated earlier, regional geography must be confined to what it can do well, and not attempt the functions of systematic geography when it will be wide open to well-deserved critcism.[3] Most geographers would admit that the two branches are complementary, without necessarily ever using both methods to do one job, but they could be even more complementary, even more useful as two different tools which will do different parts of one job, if the differences between their aims and capabilities are further emphasised and clearly defined. Thus in contrast to the aims and capabilities of systematic geography suggested above, regional geography is qualitative, literary, complex, academic and subjective. Such adjectives might appear enough to condemn a subject out of hand in the second half of the twentieth century but these are the qualities which must be stressed and used where they are appropriate. A regional geography which seeks to describe a landscape and bring out the character of an area is not concerned with exact quantities.

The writer may have to study statistics while making his analysis, but figures will not give a picture of the surface nor an idea of the way of life, and the writer will have to rely on well-chosen words to get these qualities across. One method here is to choose and describe a small area of countryside, one village or town, in order to give an impression of the whole, rather than to list and enumerate everything. This implies selection, but the selection in regional geography is its essential difference from systematic geography. The latter selects horizontally, taking one topic throughout the world but separate from all other topics in any one place, while regional geography selects one place and separates a multitude of topics in that place from similar phenomena in surrounding areas. Thus within the area or region, the study may be most complex, depending upon the intricacy of the interrelationships between the phenomena in that region. For example, the interrelationships are regarded as relatively simple in the Tundra, the Sahara and Australia, so these regions are taught to our younger and weaker schoolchildren. The north-east USA, Greater London and other large, urban, industrial areas are quite correctly regarded as so complex that it seems often that graduates have not tackled them fully.

None of the descriptions would necessarily be exhaustive nor would the information be in any familiar order for easy reference. This half of the work can and should be left to a systematic account which can fill in the gaps and provide what is considered essential factual information in a standardised order. Up to the present time regional and systematic, or special and general geography have been recognised as the two major approaches to the subject, and apart from the works of the ancient giants Varenius, Humboldt and Ritter the tendency has been to keep these separate, in different books, even with authors specialising in one or the other, and they have been truly complementary only if students have taken the trouble to obtain both regional and systematic works and read them both for a given continent. Knowing that this is not very likely, and even less likely for the general reader, the tendency of regional geographers is to be as complete and orderly as possible, and this weakens the impact of the regional portrait. The contrast is that between a portrait and a police record. If an artist must show full face *and* profile, and give the particulars of height, weight, colour, age, name, occupation, address and so on in a rigid order, then this insistence on completeness and order will detract from the portrait. Whereas if the artist may choose his pose, viewpoint, lighting, palette and detail, he is best able to bring out the person's character within the limits of his medium. Neither the painting nor the photograph and record card are complete in themselves; together they give a fuller picture which may be filled out even more. Similarly it would be better for the geographer to remain faithful to the regional monograph and add to it a systematic account in the same book, rather than to try to make the regional description a systematic geography at the same time and hence weaken both pieces of work.

There has been some development in this direction but a distinction must be made between those works in which a systematic section is simply an extended account of the physical features as in Mutton's *Central Europe*, or an economic survey to support the regional account as in Ormsby's *France* and Walker's *Italy*, and those works where the systematic account is balanced against and complementary to the regional portraits, for instance Spate's *India and Pakistan*. Another method is illustrated in James's *Latin America* where certain facts are set out in note-form at the beginning of each regional chapter. These are disappointingly brief and certain facts, such as export figures, cannot be found anywhere in some of the chapters, but the idea is sound and capable of further development. It is significant that the facts of area, population, birth rate, labour force, leading crops and so on are put separately at the beginning of each region, and not all together in an appendix at the back, emphasising that these facts are presented thus for information in the most suitable form and are not just statistics to back up statements and arguments in the text.[4] This method of chapter heading could be developed into a lengthy section for each region where all numerical facts are tabulated for easy reference and where essays on isolated topics such as geology and wild vegetation (which in a particular region may have no significant effect on the rest of the geography) may be included for completeness. If the lead given by James is followed, then regional geography and systematic geography will be complementary in fact as well as in theory. Moreover they will be complementary within one book, and, ideally, within one chapter.

This is not a case of advocating new types of geographical works, but a case, rather, of calling attention to types which already exist which need more exact definition and which are capable of much greater development and use. The two types of geographical work mentioned so far, regional and systematic, have been described and defined in many other books,[5] but it is clear that other variations may be fairly distinguished. The old-fashioned, rare, regional monographs of the French geographers can be contrasted with the descriptions of continents on a regional basis which are now being written all over the world. The former, which did not flourish and spread, were much nearer the true spirit of regional geography than are the latter. The work sprang from an interest in one region and was an attempt to convey the unique quality of that region by choice and arrangement of information about it. At the other extreme, the description of a country or continent on a regional basis springs from a much less intense and more academic interest and is simply a method of dividing an area into the least number of sub-areas in order to be able to generalise about them. Within each area, or so-called region, the information is actually given in a systematic way. There is no selection of the phenomena to be described and usually very little rearrangement for the most effective presentation. To one not very familiar with geography, the usual range and order of phenomena may seem at once arbitrary in choice, and at the same time logical in order, but too much geography is written with a slavish, uncritical acceptance of the list of phenomena to be considered and no attempt to change the order for the best effect. Strict adherence to the list may be justified for completeness, but there is little justification for blind obedience to an order of working which at best is pedantic and at its worst leads to plain determinism.

Unintentional overlapping of systematic and regional geography is much more common than its conscious complementary use. In addition to the fact that modern regional geography degenerates into a systematic account, much ostensibly systematic geography has a strong regional component. Works of physical geography are less liable to this, concentrating on general principles of geology, geomorphology or soil development and simply selecting regional examples. Climatology tends to be more exhaustive and follow general considerations of elements and factors of climate with a complete description of world climates, but then the possibility and desirability of defining climatic regions is greater than that of defining regions in the other phenomena. But some works which at first glance appear to be systematic, economic, political, social or human geographies turn out in fact to be merely world geographies with the emphasis on economics, all Man's activities, or on political units. These are just larger, more generalised accounts of the world than the continental geographies, again on a regional basis. For example, political geography takes the nation state as the essential unit, the region, uses the national boundaries as the perfect definition of the region, and puts more emphasis on the facts of shape, size and position of the region than does ordinary regional geography.[6] However, the main reason for the merging of potentially systematic human geography into world regional geography is its inability to form general laws and principles. As Hartshorne, and Wooldridge and East have pointed out, the very subject matter of human geography, with its diversity and unpredictable human behaviour which contrast with the uniformity and probability of natural phenomena, prevents any valid generalisations which will apply at any time in any place. Apart from a few general economic laws, which derive from the study of economics, and a few statements about human behaviour which are so superficial as to be worthless, all human geography must be specific, about actual people in a particular place, and this is Special Geography in the correct sense of the term.

While much systematic geography may arrange its material on a regional basis, if only by continents and countries, it is in fact still systematic in that it is limited to one major topic and considers that throughout the world. But we see that in economic, political and social geography the subject is unique in, and peculiar to, each part of the world, that there has been little success in formulating general laws, and geography has been limited to describing special cases. This does not matter, as the chances of applying general laws worked out on earth to some other inhabited planet are very remote; but it does mean that we ought to distinguish between systematic physical geography and systematic human geography more clearly. The former is nomothetic, forms generic concepts, and is in fact General Geography. The latter is idiographic, dealing with specific cases and in a sense is Special Geography without necessarily being regional geography in the fullest sense of a complete synthesis of all relevant phenomena in one area. Rather than reinforce a failure by trying to wring generic concepts from human geography (a task which Hartshorne considers fruitless) it would seem better to adapt to the nature of the problem and follow up the success of regional geography in presenting the earth's surface as it actually is. The aims most likely to bring success are to use the regional description fully where it is apposite; to augment and complement this with systematic accounts; and to stop criticising regional geography for failing to do what it is not designed and never intended to do. To this end, it may be useful to put things in perspective by listing the regional description as one of several variations of method which have in fact been used, whether or not these methods were consciously formulated exactly in this way. Two definitions are necessary first, to guard against possible misunderstanding. The word 'systematic' has been used so far, as it is used throughout geography, as a synonym for 'general' geography, but in this list it is used in its exact sense of 'arranged on a regular plan'. Secondly, regional geography here means any description divided on a regional basis.

Only two types of geography have been clearly defined, although several others exist in fact. These two types go under various names but essentially the difference is between the horizontal selection of phenomena which considers one or more topics throughout the world, and the vertical selection which considers many (but not all) phenomena in one region. These may be tabulated as follows:

Horizontal selection	*Vertical selection*
Geography of topics World-wide, generic, nomothetic and usually called General or Systematic Geography	Geography of places Continents or smaller areas, which is specific and usually called Special or Regional Geography

Special, regional geography may be written in a systematic way, that is to say, working steadily through the regular plan of rock, relief, climate, soil, etc. Therefore both General and Special Geography can be systematic, and Special Geography may also be treated in a peculiar, individual way for each region when it may be known as a compage.[7]

However, geographers have not been constrained by this theoretical division and the following variations and combinations may be found. Two points are to be emphasised: that regional description is just one of several methods, and that it will be strengthened by making fuller use of the other possibilities where they are more appropriate.

1. General geography of one or more topics. Usually of a physical topic such as geomorphology or climatology but sometimes of economic activities or population.

2. General geography of human phenomena on a regional basis. These are the economic, commercial and social geographies which set out to be general analytical works but in fact are often regional and descriptive.

3. Special geography of a continent or a country. The whole area is divided into regions and all the usual 'geographic' topics are considered, but usually in the systematic order given in the table on p. 81.

4. In contrast to 2 above, are a few books in a sense regional in that they deal with only part of the world, but which treat that area

in a purely general way. For example *Southwest Pacific*, K. B. Cumberland (Methuen), ch. 3, 'Australia'.

5. Special geography with a special selection of topics dealt with in the order most appropriate to each region. Very rare, see the works of Derwent Whittlesey.

6. One book with general, then special, sections. Many 'regional' geographies treat the relief, climate, soils and vegetation of the country or continent as a whole, then deal with the human phenomena by regions. Sometimes a concluding economic summary.

7. One book with a combination of the special and general methods in each chapter. Rare, but see *Latin America* by James and *The British Isles* by Dury.

With such a variety of types of geography there is no need for the regional description to be systematic and exhaustive. This need would arise if it were the only type of geography, when it would necessarily have to contain all the expected information about every region. In that case it would have to be systematic and complete for easy and satisfactory reference. The regional geographer need not be distracted by such considerations from getting across the character of each region by selection and arrangement of information which seems best to him for the purpose. For systematic listing of facts one may turn to the general geographies, which are available as separate and distinct works.

The suggestion that geography should stop wooing science and re-establish its family ties with history is made not just in order that criticism may be avoided, but also because this will strengthen and free regional geography. Certain affinities with history must be reaffirmed, viz:

1. that regions are often only a means to an end;
2. that the regional division is often personal;
3. that boundaries are ad hoc and arbitrary;
4. that the choice of phenomena under consideration is personal and depends on what one finds at that place; and
5. that regional geography must be literature: it must be readable.

Geographers have only themselves to blame for drifting into this difficult position. Too often they have stated: 'The regions are . . .' or 'The area is divided into seven regions . . .', rather than stating: 'My regional division is . . .', or 'A possible regional division is . . .', or 'For the present purpose the area will be divided into the following

TABLE 2

GENERAL AND REGIONAL GEOGRAPHY: AN EXAMPLE OF EACH

REGIONS

TOPICS	N. America	S. America	Europe	Africa	Asia	Australasia
Geology						
Relief						
Climate						
Soils						
Vegetation						
Fishing						
Forestry						
Farming						
Mining						
Manufacturing						
Towns						
Communications						
Crime						
Population						
Frontiers						
Disease						

/// A general geography dealing systematically with climate throughout the world

≡ A regional geography dealing with all topics in each region of South America

regions on the basis of . . .'. The most difficult statement of all, but perhaps the most realistic, would be something along these lines: 'Regions A, B and C are self-evident, and I suggest a division of the rest of the area into the following. . . .'

If geography must forever lack the precision of a natural science, it need not lack the personal point of view and inspiration which is the life spark of history. Historians are constantly preoccupied with the problem of personal bias in their work: they try to reduce it but at the same time accept it as inevitable, just as they realise that their interests, enthusiasms and personal points of view give point to their work and arouse the interest of others. The attempt to be objective and impersonal in geography has, however, merely ended in squeamishness. An historian may approve or disapprove of a movement or a famous person but the geographer-cum-pseudo-scientist has become objective to the point of indifference, and to the point where he avoids any controversial topics. Consider the following selection of problems which are vital to the countries concerned, and often to many other countries as well:

Canada. The French demand for an independent Quebec.

U.S.A. Racial strife. The problem of the extremes of wealth and poverty side by side, with one-fifth of the population at subsistence level in the world's richest country.

South America. The façade of nation states hiding the unbalanced economic, social and political development.

Africa. The legacy of the Scramble for Africa. Apartheid. Partnership. The façade of nation states so-called where nations seldom exist.

Europe. Making the Common Market work, in effect dealing with the legacy of separate industrial revolutions which took place behind political barriers.

The coincidence of the Catholic Church and poverty in such countries as the Irish Republic, Spain and Italy.

Overcrowding in a technically advanced society.

Asia. Dangers to the rest of the world of Communism, over-population, Asian industrialisation.

Australia. The White Australia policy. Suicide.

New Zealand. Socialism. Economic dependence on Britain.

These are pressing problems which are an inextricable part of the geography of the areas, in some ways resulting from the past geography, and certainly affecting the present and future geography so profoundly that the whole essence is lost if they are ignored. While so vital, they are certainly open to argument and, fortunately, are ephemeral, so the two arguments against their inclusion in geography are that they are merely temporary aberrations on the basic features, and that they are the concern of sociology, economics, politics rather than geography. Neither of these two arguments is very strong. Geographical writings are ephemeral at the best of times; the facts in even the most recent work are out of date before publication, and Man's activities are developing and changing so rapidly that descriptions are not keeping pace. The other argument, that a certain topic is properly in the province of some discipline other than geography, is as familiar as the rejoinder that *any* topic is rightly geography if it helps to describe, explain or characterise part of the earth's surface. There is no question of historians avoiding the controversial features of the past, even though these are now settled once and for all for the people physically concerned. In fact historians seek them out, open them up again for debate, and generally revel in them. Crucial issues of the past are as interesting and informative as the more detached narrative, yet imply no decline from the highest standards of scholarship, providing the writer states both sides of the argument as fully and as fairly as he can. Indeed, historians go even further when they consider to what extent some person succeeded in his aims. While it may not be stated so bluntly, these considerations may lead to the conclusion that the person in question was a failure. People, now powerless in their graves, have been labelled tyrants, despots and fools. Thus part of the geographer's squeamishness may be attributed to the fact that the people involved in the problems mentioned above are still alive but, provided he states both sides of any argument, and analyses the problem objectively, the stark immediacy and reality should not deter him. The natural scientist correctly avoids any human aptitude, failing, emotion or interference in his work. Geographers who hero-worship the scientist ignore the most vital elements in their field of study and often seem to have reached the stage where they are no longer capable of dealing with controversial topics. For example compare how Wellington[8] deals fully with poor whites in South Africa, and avoids apartheid; or compare how Cole and Walker deal with the distribution of wealth in Italy.[9,10]

Following the lead of the historian's healthy interest in people, we might in human geography consider to what extent the following schemes have succeeded:

The attempt to make one nation in Canada.
Free enterprise in the United States of America.
Nationalism in South America, Africa and the Middle East.
New economic theories in Europe and Russia.
The division of India into Hindu and Moslem states.

These topics do merge with other subjects such as history, or they may be labelled current affairs, social studies, or some other names which are even more vague than geography, but it is significant that while they are essential to the student working for the geography scholarship examination and to the undergraduate they are not to be found in standard geography texts. The student has to turn to magazines and to books in other disciplines for information and intelligent opinion. The personal point of view is essential to regional geography and must be its strength, not covered up as an inevitable weakness. Fully developed, as in history, it can make the subject boldly humanistic, in contrast to its present position as an emasculated science.

[1] Hartshorne, op. cit.

[2] Wooldridge, S. W., and East, W. G., *The Spirit and Purpose of Geography*, Hutchinson, 2nd edn., 1961

[3] 'This work makes no attempt to be encyclopaedic' states J. M. Houston on p. 1 of his 800-page *The Western Mediterranean World*, Longmans, 1964

[4] See also Dury, G. H., *The British Isles*, for a similar method.

[5] Hartshorne, op. cit.; Wooldridge and East, op. cit.; Freeman, T. W., *A Hundred Years of Geography*, Duckworth, 1961

[6] For examples of good systematic general works see: *World Sea Fisheries*, R. Morgan, Methuen, 1956; *A Geography of Manufacturing*, E. W. Miller, Prentice-Hall, 1962; *The Geography of Towns*, A. E. Smailes, Hutchinson, 5th edn., 1966

[7] Compage: 'A whole formed by the compaction or juncture of parts. A complex structure' OED. See below, pp. 133 and 134

[8] Wellington, op. cit.

[9] Cole, J. P., *Italy*, Chatto and Windus, 1964

[10] Walker, D. S., *A Geography of Italy*, Methuen, 2nd ed., 1967

7

ALTERNATIVES TO AN INADEQUATE CONCEPT

Since 1951 most discussions about the region have referred to Kimble's criticism of the regional concept.[1] Other critics refer to it as the final condemnation, while even advocates of the regional method refer to it as if it might confound their arguments after all. The title 'The Inadequacy of the Regional Concept' implies that the concept is completely useless, because Kimble has said so. Close scrutiny of the article, however, leads to no cut and dried conclusions, but gives a fascinating insight into the views other geographers have of the region.

Kimble starts by saying that most geographers believe that regions exist but that there is dissatisfaction with the concept, mainly because it is eighteenth century and old-fashioned. Natural regions are obvious nonsense while cultural regions with 'functional coherence' and 'dynamic homogeneity' are meaningless. Having stated that he is going on to consider the use of the concept in current literature, Kimble does not do this but makes the three criticisms that regional boundaries are hard to define; that no two single-topic regions coincide exactly; and that Hartshorne's idea of the cultural region is no more definite than the older idea of the natural region, particularly as such regions are not homogeneous.

Then Kimble switches his attack to geographers for often lacking the ability fully to understand, describe and explain a small part of the earth's surface. Returning to geography rather than geographers, Kimble reasons as follows: areas are interdependent; regional unity in one area cannot exist by the side of regional disunity in another area, and as there are obviously many disorganised areas in the world today therefore there can be no unified regions. In trying to show that the regional concept is a product of the static society of eighteenth-century Europe, Kimble admits that there were regions in Europe, that the longer they lasted, the more well defined they became and that they were not interdependent, but in fact were destroyed when the industrial revolution made necessary a wider interdependence. He defines regions as historical products of the self-sufficient feudal units of medieval Europe which have been ruined by technological change. Therefore, the argument goes, we can not expect to find regions outside Europe. For example, Europeans migrating to America mixed together so much that their regional differences disappeared.

For his conclusion, Kimble returns to attack the geographer by saying that the work of description and analysis in any area is too great and therefore the only possible geography is systematic (general) geography. In passing, he states flatly that regionalism aims to make geography a science, and that people should not establish city-regions.

Thus Kimble implies that the regional concept is inadequate in the following ways:

1. It applies only to past geography.
2. It is indefinite, especially about boundaries.
3. It applies only to homogeneous, static, isolated regions.
4. It can apply only to Europe.
5. It presents an impossible task to the would-be regional geographer.

It is hard to believe that Kimble can imply these limitations when so much modern geography describes functional regions all over the world with some precision and a certain success, until one recognises that he has toppled only one tin soldier, but one which may in turn topple all the rest. For Kimble has taken one very personal definition of the region, shown this to be inadequate, and left it open for others to leap to the conclusion that therefore *any* definition of the region is inadequate. What complicates the issue is that he does not define the type of region he had in mind while making the criticism, but this can be determined by examining his arguments very carefully. It becomes clear that for the purposes of that particular chapter, Kimble's concept of the region was that it is:

1. composed only of human, cultural or sociological phenomena, 'a medal struck in the likeness of a people', physical phenomena being excluded;

2. a formal, homogeneous area with functional unity (a virtual impossibility which the present writer has not encountered in any other work);
3. self-evident on the ground;
4. perfectly static, and
5. an area where the boundaries of each phenomenon under consideration coincide perfectly with all other boundaries.

The most obvious point is that in the majority of regional works, physical phenomena, especially relief and climate, are considered at least as important as human phenomena. Formal, composite and functional regions, with a possible hierarchy, have been recognised. Some self-evident regions may exist but where a geographer is using the regional method as a means to an end we accept that some divisions are the results of his analysis of maps, and are not to be found on the ground. It is surprising that Kimble insists on static regions when he seems to put so much emphasis on the historical viewpoint from feudal times, through the eighteenth century to the present state of the developed world. The phenomena studied in economic and social geography change constantly, and ever more rapidly so that any pattern observed at one time is fluid, and cultural regions and human geography must be regarded as dynamic. The dynamic industrial societies of the nineteenth century produced well-defined industrial regions in Britain, Europe and North America. But this pattern of regions is now obscured by the mushroom growth since then of an increasing variety of 'light' and 'footloose' industries. Governments with their development areas and regional plans are as far behind the dynamic reality of private development as are some geographers. Finally, many regional works imply that the boundaries of climatic, soil, vegetation and land-use zones coincide with each other, but in fact these boundaries are so vague as to leave much leeway and in fact, as stated earlier, divisions of relief usually form the regional framework and very rarely is it claimed that boundaries do coincide.

Once these points have been disentangled from the argument it is obvious both that few other geographers think of the region in this way and that such a concept is very easy to criticise. An intriguing point here is whether Kimble had these points clearly before him when he started. If so, he may well have defined the region in these odd terms simply to provoke argument and further thought. If not, and Kimble, like most other geographers, felt that the regional concept was so well known as to need no preliminary definition, then the present work has even more point in that, besides attempting to set down all the ideas about the region, it may make others aware that their concept is vastly different from the next man's.

Whether or not Kimble meant to provoke argument, he must be taken up on several points, and it is emphasised again here that the present criticisms are of his full argument, and not of the summary given above. Firstly, his ideas of form and function are very confusing. He puts great emphasis on the appearance of the region, that it must be self-evident, look different from its neighbours and even be discernible from the air. This he calls a formal region, but emphasises that it is self-sufficient and isolated, basing his ideas on the feudal village and the eighteenth-century European countries. But a self-sufficient isolated community is not formal and homogeneous, it must be varied and have different parts with different functions in order to survive. In fact, at one point he complains that large 'regions' such as the Corn Belt and Cotton Belt are not uniform. The main contradiction lies in two separate attacks on the concept, one along the lines that as all regions are interdependent, if one fails they must all fail (economically), therefore regions are bound to collapse completely; the other is along the lines that following the eighteenth-century pattern regions should be isolated, and as they are not, regions no longer exist. Thus in separate places he is denying and asserting the existence of interdependent regions.

Secondly, Kimble denies the possibility of regions outside Europe, and his concept of regions composed exclusively of human phenomena is fundamental to this point. Again his peculiar views both of the European type of region and of the landscapes overseas are responsible for this denial. Having emphasised that regions must be home products, natural growths like zonal soils, he states that there can be no regions in North America both because the people are new arrivals, 'foreigners' on that landscape, and because Norwegians, Hungarians and Italians all became Americans. One may agree that European differences have been lost and that the European type of region is not to be expected in North America. But either Kimble's prejudice against anything non-European or his lack of detailed knowledge of some other parts of the world prevent him from admitting that while we lost Europeans in the melting pot, we now have equally distinct regional types in the moulds of the South, the Midwest, California, New England and so on. Can Kimble really insist that these are not recognisable regions, even if they are not the type of which he approves? There were no regions on the frontier, as he says, but the frontier disappeared long ago and differences have had time to develop in North America on a larger scale than in Europe.

This particular argument of Kimble's would have been much stronger if he had taken Australia as his example. Of all the new lands settled by Europeans Australia is the most difficult to divide satisfactorily for the purposes of description. Regions in the European sense certainly are not self-evident, the three main features of relief, land use and political divisions cut across each other in a most unusual way, and it is necessary to conceive a basis of division peculiar to that continent, in a sense to devise an Australian type of region. Thus Kimble has not proved that the regional concept is inadequate outside Europe; only that his special concept does not apply. Most geographers would not expect to find replicas of the Highlands, the Paris Basin or Basilicata all over the New World. Moreover, they would be most disappointed if they did, for the interest and purpose of geography lie in the fact that each part of the earth's surface is unique and that in a normal lifetime with ordinary faculties and resources there is always something new, different and interesting to see and learn. One might argue that there would still be the possibility of different regional characteristics within the framework of formal, static, self-evident, clearly bounded units, but the necessity to enlarge and adapt the regional concept to work in all parts of the world has proved to appeal to geographers much more than has the convenience and safety of one ready-made rigid concept peculiar to Europe.

Other personal attitudes are revealed in Kimble's argument which, while not strictly concerned with regions, may help to explain his attitude to the concept. He seems to believe that all geographers are searching for a standard model region and states that geography will be important only when all geographers can agree on such a thing. Less clearly stated are the points that a region is organic (which conflicts with the more definitely made points that the region is static and formal, but is another contradiction which points to the emotional nature of this criticism) and is used by the geographer to prove the workings of determinism. As Kimble, like many others, can easily refute determinism, this is one of the attacks on the regional concept which seems to succeed, superficially. More clearly, however, he states that geography must be useful, and goes on to tell geographers what they ought to do, what their subject ought to be. He advocates systematic study of topics of general geography including conservation, settlement, dwelling patterns, house types, field patterns, disease, pioneer settlement and marginal areas as well as the more usual familiar ones. This is most dangerous ground for anybody. We all study what we want to study, not what someone else thinks we ought to do. The motive is interest rather than duty, and quite as many geographers are interested in regional contrasts as in world-wide distribution of one phenomenon. If our interests lead us far beyond what is conventionally recognised as the boundaries of geography, then the only possible demand can be that we stop calling ourselves geographers, not that we drop our interests, curtail our studies and come back into the fold. In so far as geography is an applied science, then geographers may want it to be important, useful and untainted by the human element, and some director of research might pertinently indicate fields in which research is vital, regardless of personal interest.

The most disturbing point in Kimble's criticism is the suggestion that regional description is an impossible task. He believes that the work of collecting all the necessary information about a region, analysing it, and writing a correct description is too much for one man. The size of the region in question, the number of topics to be considered, and the amount of detail are not specified, but any student is familiar with the feeling, which disappears as the investigation is pressed forward, of being overwhelmed by the amount of information. He argues, therefore, that the work should be split up among several people, in effect that regional geography should be destroyed by being dismembered. This argument, however, depends on one's view of the purpose of geography. For Kimble, in this essay, it must be useful and definite; but the present writer regards the purpose of geography as purely intellectual, academic description for world comprehension, and for this end regional geography of areas which by nature are ill-defined is essential. One cannot show the regional concept to be inadequate by making the regional geographer feel inadequate. Using exactly the same argument one can make the general geographer feel inadequate by stating that one man cannot analyse and describe all the climatic variations of the world, or all its economic activities. The trick is to avoid defining the size of the problem. If the general geographer can limit himself to one topic, in a specified amount of detail, so the regional geographer can limit the size and number of his regions, and the choice of topics within those regions to what he can comprehend and handle.

To a certain extent, if one man cannot write the geography of a region, one reader cannot understand what it has taken several men to write. The secret lies in how much original work one man has to do. If geography is regarded as a practical subject, then the work *is* impossible for one man. He would have to research, tabulate and map everything from rocks to population before he could even start to analyse and write. A geographer may need to be able to make geological, topographical and soil surveys, to collect original data on occupations or to take a census, in case in some part of the world information on one of these topics is unobtainable any other way. If no maps and statistics are available at all for an area, then academic geography is impossible until such time as they are, and anyone setting himself the task of collecting all the information is trying to be a surveyor, geologist, geomorphologist, meteorologist, social scientist, census recorder, demographer, etc., all at the same time and is obviously committed to inadequacy and failure. Practical knowledge of these people's work is essential to fill a gap left by one of them, but not to do the work for them all. If the geographer has to do all the original work in these fields and then start on his own work, the result, if he survives, is likely to be shallow and useless.

In contrast, by admitting from the start that regional geography, at least, is not practical work, the geographer puts himself in a much stronger position. He is then free to take the results of other experts and build up his composite picture. This is the only way to make progress, to start where others leave off. Many other disciplines provide the raw material for regional geography, so that at once the single geographer has the conclusions of many experts in many fields. Hartshorne[2] and others have gone fully into the unique position of academic geography as a cross-section of many other disciplines, but the emphasis here is not so much that geography cuts at right angles to, let us say, geology and economics, but that it takes the results of these studies, results which may have been the life work of many other people. The geographer's justification is that he is interested to combine and correlate subject matter widely separated for the purposes of intensive study but vitally interconnected on the earth's surface. He is doing, in fact, what Kimble implies the reader ought to do. Kimble's argument is: regional description is too much for one geographer, therefore each topic should be treated by an expert. One assumes that the end product is a volume with each chapter written by a geologist, climatologist, etc., and that someone is going to read the book and be able to comprehend the whole. If so, the regional geographer has a similar function to the intelligent reader, plus his urge to analyse this material, combine it, rearrange it and present it as his regional synthesis. The geographer's purpose may be to satisfy his own curiosity but in the process he has the function of medium between many investigators and the lay reader. The end product of one man's work may not be as perfect and detailed as a symposium, but more important than perfection beyond comprehension is the anxiety and the attempt to know. The work may then be limited by one man's ability, but provided he is well practised in his work, and avoids useless repetition of work already done, then the limitations of a writer are not necessarily greater than the limitations of his readers. In fact his driving interest and professional practice are likely to make him a more competent guide, for while further analysis is often necessary in regional work, the essential process is the final synthesis of an idealisation of reality.

The more one studies Kimble's attack the more one suspects an ulterior motive. It has the function of forcing a reconsideration of the many points raised, and in the process one has to define one's own ideas in order to defend them. Thus the importance of the attack is not whether it succeeds in showing the inadequacy of the concept but that it demands clarification of the subject on both sides. Of the many valuable lessons to be learned from all this, the most important seems to be that too hard and fast a definition of the region or the regional method would be ridiculous and prevent its universal application.

For the moment, however, it may be fruitful to imagine that Kimble's or some other's criticism had shown the concept to be inadequate, and to search for an alternative. Even if the most open-minded search fails to suggest anything more satisfactory, some other methods may prove to be useful at times. The only assumptions to be made, then, are that some division of the whole world is necessary to provide areas of a size convenient for description at a given level of intensity, and that this is for the purpose of vertical study of many topics at once, as in Special Geography. One partial conclusion of this whole work so far, is that while in some places distinct, recognisable 'regions' do exist, as the English Fenland and Lake District, this cannot be said of the whole land surface of the world. For these less distinct areas, however, the regional method has been used with considerable success. Therefore the alternatives will be considered from the two different points of view:

A. alternatives to the regional method
B. alternatives to the region.

A. Alternatives to the regional method

1. Instead of trying to find boundaries in gradually changing phenomena in order to construct second-class regions, one might

take units which are universally recognised in all walks of life, continents, islands, countries, states, counties, parishes and so on but not units such as lowlands or mountain chains about which there can be argument. The changing basis from natural to political units is no drawback providing the criterion of indisputable, universally agreed units is maintained. In the Old World the boundaries of the smaller political units will probably fit the boundaries of any one other phenomenon as well as, say, land use and relief regions coincide now; but in the New World, especially in the rectilinear states of the American west and in Australia, obviously not.

2. Divide the surface only by lines of latitude and longitude, to the nearest second if necessary, but using only north-south and east-west lines as in the present convention. This might work well in Chile, Algeria, or Java, but not in the Appalachians, Jura or New Zealand. The idea has the appeal of infinite choice of boundary lines combined with precise definition, but units would tend to be as artificial as the pictures made up of coloured squares on an embroidered sampler. A similar system has to be used for the division of topographical maps into sheets of manageable size, but most people are aware how too rigid an adherence to the national grid in the 1:25,000 sheets and to one size of sheet in the 1:63,360 series leads to division across the centre of some particularly interesting or significant area.

3. The International one-to-one million sheets might form the framework at one appropriate level of detail with the double recommendation of a single criterion for division and one that is known and accepted throughout the world. The drawbacks are the rigid framework and the fact that it exists only at one scale.

4. In order to reduce the drawbacks of shape and size inherent in the above method, a conventional system of hexagons might be applied to the surface of the globe, just as they cover the rotund surfaces of such fossils as Lithostrotion, Cidaris and Holaster.[3] As suggested in Chapter 1, this is the shape of region preferred, while subdivision to produce similar shapes at any desired size is possible, as Christaller has shown in his idealised diagram of urban fields.[4]

5. The search for real, permanent or incontestable units brings one finally, and rather quickly, back to the old idea of drainage basins. These are suggested again here, not with any implication that they are 'cradles of civilisations', but because they can be mapped and are among the most permanent features of the earth's surface. The main drawback is their size, but study of such a map as that in Bartholomew's *Advanced Atlas*[5] suggests there are certain

advantages, particularly: an eastern boundary for Europe; inclusion of North Africa with Europe; tripartite division of Africa; India and Burma balanced against China and South East Asia; the significance of the Australian coastlands, and so on.

B. *Alternatives to the region*

The obvious contrasts between the deserts, temperate regions and frozen wastes, known since the days of ancient Greece, perhaps lead us too far in the direction of looking for sudden changes between them. One might hope for regions as clearly defined as the different fields of crops in a European arable farming area. Perhaps, in fact, we ought to be concentrating on the gradual, imperceptible change from one association of phenomena to another, and this will be considered below. Whichever is the case, real units, as suitable alternatives to the ever-elusive regions, are hard to find.

1. The political unit, from the state down to the parish, is the most obvious. At the level of the nation state it is used perfectly successfully by political geography, but the method is not pursued ruthlessly through the constituent counties, pays, cantons, wards, etc., in spite of the attractions of shape and size[6] and the fact that most statistics of use to the geographer are made available on the basis of these administrative units.

2. As an alternative to the multiple-topic region, the single-topic region is more easily defined and more readily acceptable. If this is to be brought into wide use, however, it will need a different proper name. All other phenomena can be discussed within the regions defined by one phenomenon, but some bases are better than others. Trying to decide on one which fits other phenomena most closely, or has most influence on others, begs the whole question and prejudices the rest of the study. Such a decision can properly be made only *after* all the work is finished. Therefore the best criterion is that the unit should be visible, the smaller ones on the ground, the larger ones on topographical maps which represent only real, tangible objects.

Thus:

(i) Again relief forces itself to the attention as the one feature of the physical environment, directly, universally and instantaneously observable.

(ii) It is doubtful whether much genuine, original, natural vegetation now exists, but even in Britain vegetation covers the greater part of the landscape, and we have units of different types of surface. By combining natural vegetation, wild vegetation, cultivated vegetation (i.e. the *actual* vegetation in every place) with bare rock, bare debris, bare soil, water, ice and buildings, generalisations on the actual surface, say observable from an aeroplane on a clear day, can be made. Units of vastly different sizes are thus at once revealed by their texture. Moreover the very fact that this one idea of texture is composed of the factors of microclimate, of soils and of units of land use, establishes that there must be a fair degree of correlation with climate, soils and economic activities.

(iii) The only other features with all the qualities demanded are Man's constructions on the landscape, fields, settlements and communications. Regions of certain field patterns, farm and house types, rural settlement, urban sprawl have been mapped[7] and their characteristics are in no doubt, even if their origins are. But the surfaces of the continents are not completely covered by these features and so settlement zones can be of only secondary importance.

These possible alternatives to the region have drawbacks as serious as the drawbacks of the region itself. Political units involve repetition where they are too small and subdivision by some other criterion where even the smallest is too large to coincide with, say, an area of land use. Relief units, suggested here precisely as relief units and not in their usual disguise as the framework of multiple-topic regions, always present the problem of where to draw the line. Should this be at the base of the mountains, along the watershed, or along the river? Rivers have at times been considered the backbones of communities, but the larger, wider rivers were more often an obstacle than a unifying factor before the days of steel bridges. So either much variation in demarcating relief regions, or unnecessary subdivision of other phenomena, is inevitable in the use of relief units. It is suggested that textural units have more significance for the study of many topics in the same area, but at the same time there is no doubting the wide range in the sizes of such units from a polar ice cap to an English orchard or the relative homogeneity of the ice cap in comparison with the mosaic of juxtaposed surfaces say from Denmark across the Alps to peninsular Italy. The last suggestion, of cultural units, while admittedly of limited use, is put forward quite seriously. It may well become of greater importance both as Man is forced to extend and intensify the settled areas of the globe, and as Professor Hoskins' systematic study of topography, combining points (i) to (iii) above, progresses as a separate discipline.

The only other possible alternative to both the method and concept of regions is to go to the other extreme and to reject boundaries and definite units completely. This is not to deny regional differentiation or the possibility of correlation between different phenomena, but one way round the argument about the existence of regions and their delineation is to transfer the emphasis and attention from margins to centres. The methods necessary for this are those of the presentation of facts rather than their subdivision, and several variations come to mind:

1. For any area, describe each element or topic of geography in turn, systematically, and finish the work without any attempt at subdivision or correlation.[8]
2. For a selected area map and describe the distribution of every phenomenon which varies from one place to another and for which sufficient information is available.
3. Continue to present the facts of any and every topic by dot maps, isopleth maps or choropleth maps which have no boundaries other than those found in real life. This calls for considerable skill in the choice of the value of each dot, the exact placing, the choice of isopleth values and the size of the steps in choropleth shading[9] but it is possible to represent the various gradients of reality rather than to replace these gradients by arbitrary lines.
4. Taking method number three to the extreme, each element of the geography of a continent might be presented by means of maps, diagrams and aerial photographs alone. This is certainly an extreme suggestion, but it is made with the realisation that method number one above is so difficult as to be practically impossible. One may use words alone to describe the change in phenomena in one direction easily enough; but to describe phenomena changing in every direction from many points is a different problem, yet one solved very easily by means of the map.

All these suggestions are variations on the theme of taking each element of geography from structure to population and describing its spatial variations by means of words, maps or photographs. It is implied that there are considerable spatial variations in at least some of these elements but that nowhere does the geographer (a) draw artificial boundaries or (b) attempt any correlations which might distort the presentation of the facts. These two points are very important, for if they are observed it means that any reader can choose an area of any size, shape and position for his particular purpose and compile the regional geography of that unique area from the facts so presented in the most objective and realistic manner possible. This comes close to Kimble's implication that the final synthesis should be left to the reader. However, the essential point here is not that the geographer is incapable of making a regional synthesis, but that, realising that an infinite number of equally valid regional syntheses are possible, he refrains from clouding the issue with his own personal selection. The work of the geographer, in this case, lies very close to the original meaning of the word geography, for it is to produce a reference book of certain topics with the possibility of rapid and accurate cross-reference from one to another. Such books do exist, but we regard them properly as atlases rather than geography books, yet they are as different from the concept of an atlas as a book which shows where places are and how to get to them as they are from regional geography books. *The Atlas of South East Asia* (Macmillan), *The Atlas of Western Europe* (Murray), *The Faber Atlas*, *The Oxford Economic Atlas of the World* and above all *The Atlas of Britain* (Oxford) may be used in the way suggested to compile the geography of any area, selected for any special reason. As another comment on this question of who does the research and who makes the synthesis, the use of any of the above-mentioned atlases must impress the reader with the amount of time, skill, special resources and division of labour necessary to produce that much accurate, detailed information in a useful form. There are those who see the geographer's proper field covering the beginning of the work, plodding about in some remote place collecting data and making maps; Kimble would even deny him the ability to comprehend what he has collected. However, it seems to the present writer that the regional geographer's work comes towards the end of the process in that he devotes his time, energy and skill to the final synthesis. *The Atlas of Western Europe* does not show at second hand what one could see for oneself if one were not so lazy; it shows the results of original work, analysis in many spheres, distributions which are to be seen only on a map, and by one man only in an atlas.

Consideration of methods 3 and 4 outlined above, combined with study of such an atlas as *The Atlas of Britain*, raises two questions which are vital to an understanding of regional geography. The first question is 'How do maps influence the geographer?' and the second 'How was the first regional division in a given area made?' These are two quite distinct questions but the answers depend on each other to a certain extent and so consideration of both questions will be combined here although certain conclusions were worked out separately before this presentation. Any regional division can be made either by direct observation of the landscape or by interpretation of information about the landscape in the form of symbols. The first regional division of any area, therefore, may have been made by direct observation, but by the very nature of the work the size of the area so divided, or the detail and accuracy, must be severely limited. At one end of the scale a man on foot may get to know an area twenty to thirty miles across, say the Lake District or the Peak District, and by direct observation may be able to find possible limits of the region in which he starts, such as the edge of the Lancashire Plain or the Pennine Slope. At the other extreme a man with the resources to fly frequently may observe regional differences of a continent but only by means of traverses or transects. If, in the former case, he wishes to know a larger area, or in the latter case wishes to have more detail of the continent, then he must start to draw on information supplied by others, especially on maps. These are the alternatives available today, when we have countless students and aeroplanes, but knowing the historical development of geography one can say quite safely that very little of the original regional work was done from the air, and only in France were there enough students starting from different centres to give it complete first-hand coverage; therefore much of the work was done indirectly by means of maps.

The question was framed carefully so as not to assume, for the moment, that self-evident regions exist. However, regional divisions have been made and presented in the textbooks in increasing numbers over the last fifty years. More recently, maps and atlases of the type which must have been used for the original divisions have become much more widely available, and it is most instructive to compare these maps with the maps of the 'regions' in the textbooks.[10] *The Oxford Economic Atlas of the World* is perhaps the most useful in this respect, on a world scale, as the method of mapping by dots avoids the introduction of any artificial boundaries by the cartographer. Admittedly only in North America have geographers agreed in defining regions on an economic basis, but a comparison of all the economic maps shows the persistence, from one commodity to another, of two distinct sets of regions, the major world industrial regions:

East-central North America
Britain and Western Europe
USSR
Japan

and the major world agricultural regions:

Central and Eastern USA
The Pampas
The North European-Asiatic Lowland
The African Savannas
India
China
The Coastal Fringes of Australia

It is implied here that if all the separate maps of manufactured goods, crops, and animal products were compiled on to one map, these areas would be overprinted again and again, leaving the rest of the land surface almost bare. A single commodity very rarely shows such marked contrasts in distribution that one region can be separated from another. The general rule is for the mapping to reveal gradually changing densities, with no steeper gradients anywhere to mark the boundary between regions of extensive and intensive production. Moreover the dots are often so widely scattered at the fringes that even the boundary to the total area of production is difficult to define. At the same time it is perfectly possible to name places which have high or low production and to contrast these two centres, although there is no break in the gradation between them. A few exceptions stand out. Looking at maps 21 and 24[11] there can be little doubt of the existence of the Corn Belt or of certain rice-growing regions, the concentration of the dots is so great. Yet the concentration of dots on map 23 is just as great, but how famous and important is the 'Potato Belt'? Obviously the evidence of the map is not the only criterion in the delineation and naming of regions. It is always emphasised that cattle, pigs and soybeans are just as important in the Corn Belt as is corn, but corn is the regional characteristic. Maize growing ten to twelve feet high is impressive, as are the paddy fields in the rice regions; but a field of potatoes is neither impressive nor always instantly recognisable to the layman and thus the significance in the landscape may be just as important as localisation in one part of the world to the regional geographer. Other commodities worthy of mention for their localisation and therefore high value in characterising regions are Grapes (32), Olives (45), Soybeans (44), Pigs (51), Vegetable oil seeds (44-47), Sheep (54), Jute (60) and Cotton (64).

While indirect observation through maps must be used for geography on a continental or world scale it is clear from the foregoing that geographers have not always been bound by the facts presented in the most objective way by the cartographers. Judging by the evidence of these economic maps, some obvious regions have been ignored, while others, hard to see and define on the maps, have been created; or at least developed and emphasised. However, it was shown in Chapter 4 that it is not necessary to keep to one criterion in defining regions, and a set of maps showing the distributions of a wide variety of phenomena may have greater significance than a set of economic maps. *The Atlas of Britain* offers an excellent illustration of scores of maps covering every topic of the physical environment and economic activity, and some topics of the social life of this small island in considerable detail. Moreover, direct comparisons can be made between any maps by means of the overlay grid. The only drawback is that the methods of mapping some of the topics, suitable as they may be for general reference, tend to over-emphasise some areas and therefore may prejudice this rather specialised search for regions. With minor variations the standard texts on Britain are usually divided into about fourteen regions, with the Highlands, Central Lowlands, Southern Uplands, Cumbria, Lancashire, Yorkshire, Wales, the Midlands, East Anglia, south-west England, Hampshire Basin, Thames Basin and the south-east scarpland common to most.[12] Regions so persistent in the textbooks over the years ought to be discernible in some of the maps, if the regional division is based on observable phenomena. However, this is not the case with the economic maps where one might expect to find regional differences of agriculture corresponding, say, to differences in the physical environment. A few crops, such as oats, potatoes, fruit and sugar beet, do show remarkable localisation in north-east Scotland, the Fens, Kent and the eastern counties respectively; but then the much more common distribution of imperceptible gradation from zero to the maximum concentration is repeated time and again in the maps of the other crops and of the animals. The maps of wheat and barley show how difficult it would be to define regions, and how the idea of East Anglia being the cereal region of England is as much the geographer's idealisation as a true statement of fact. Consideration of such maps, however, brings us to the crux of the problem. The best way to communicate the facts of the distribution of farming is by means of maps; it is virtually impossible to convey such a complicated picture in words, and yet the final presentation is usually required in words and so must be less accurate than is possible. There is a similar, familiar, problem in photography. The best way to reproduce the colours of the original scene is by means of a transparency in a projector; yet most people prefer paper prints, of which they can have any number of copies to hand round at any time, under any circumstances, and so are content with colours less brilliant and less accurate than the transparency. An atlas such as *The Atlas of Britain* was suggested as one of the alternatives to regional geography, but by itself it is soon seen to be insufficient. An atlas plus a commentary is required, and any detailed commentary on the farming must refer to parts of the map where (say) cereals are non-existent, of moderate importance, and finally predominant, and so regional divisions are thereby implied, if not actually stated.

Still concentrating on one topic in isolation, comparison of the maps of permanent and temporary grass on pp. 80 and 81 should make the influence of such maps clear. The map of permanent grassland has steep gradients between areas of contrasting density and it would be possible to draw generally acceptable boundaries on this map. The map of temporary grassland, in addition to being a transitory basis for any division, shows very gentle gradients which offer no breaks or contrasts along which to draw boundaries. In many of the maps of farming topics, however, while it may be pointless to try to define regions, two features are so persistent as to demand description. These are the existence of Highland Britain as a 'negative' area, and the characteristic complete change in emphasis between the eastern and western sides of the country. This gradual change from one side of the country to the other, shown so well in this atlas, is familiar to all students of the geography of the British Isles. Yet in the textbooks it is usually described as a contrast between east and west rather than as an imperceptible change; the west and east of the central plain of Ireland are set forth as contrasting pastoral and arable regions; similarly the dairy farming of the English west Midlands is contrasted with the cereal growing of East Anglia as though there were some definite boundary. Even in northern Scotland and northern England where the hills interrupt the pattern, the atlas shows a more gradual change in the eastern Highlands and in the eastern Pennines than the written texts make clear. A definite line down the centre of the country and a vivid contrast between pastoral west and arable east may just be acceptable in school textbooks for junior forms, but the pretence of precision is not impressive in a supposedly advanced text. If the change of farming across the country is in real life as gradual as *The Atlas of Britain* shows, then the best geographical description of the country must resist the regional temptation and characterise this merging of the one with the other.

Returning to the other point, we note that several maps in the atlas present only two possibilities to the observer. Many of the maps of crop distributions and the maps of Land Use, Agricultural Labour and Agricultural Holdings on pp. 68–70 show the greatest contrast as that between the area where farming is predominant (Lowland Britain) and the area where farming is virtually non-existent (Highland Britain). In general geography and in the consideration of single-topic and limited-topic regions such as farming, the possibility, in fact the very existence, of areas which are of no interest under that particular topic is obvious. Thus the main division is not between different types of farming, but between the total farming area and the total non-farming area, which in Britain is particularly coherent and outstanding.

Actual inspection of the maps, then, does not reveal the regions so often drawn and described in the textbooks, nor does it suggest how, say, farming regions such as those described by Stamp and Beaver[13] were arrived at in some other way; unless they are in fact based on relief regions without the authors actually stating this. The same criticism applies to the industrial regions, for the student familiar with the contrasts drawn between the textiles of Northern Ireland, steel and shipbuilding of Scotland, hardware and cars of the English Midlands, tinplate of South Wales and so on may be bewildered to find that the atlas shows precious little difference between one industrial region and another. The only reassuring fact in this featureless plethora of plotted information is that the traditional industrial regions are still where we thought they were. For the only similarity between this up-to-date mapping and even the most recent textbooks is that all the maps of the individual industrial products do show again and again the persistence of such areas as the Clyde Valley, Tyneside, Lancashire, the West Riding, the Black Country, South Wales and Greater London. Apart from this, the only significant localisation of any industries is of the textiles, with most of the cotton industry which remains still concentrated in Lancashire, and the woollen industry still firmly in the West Riding. This illustrates a most important point about regional geography. The raw materials of facts are constantly changing, and any method of regional division, or any system of regions, must be capable of change in order to reflect this development accurately.

The maps in *The Atlas of Britain* reveal vividly the complex variety of all our industrial areas. The problems of industrial areas dependent on only one industry are put forward in our textbooks, but since geographers became aware of these regions and their problems not only have the regions varied their activities but towns have developed completely new industries, and even many single firms now embrace a number of seemingly unrelated manufactures in their search for stability and continuity. So while the position and the extent of the major industrial regions remain fixed, and while the location factors for the original industries remain the same as matters of historical record, the characters of the regions have changed completely. Now that skilled labour is less important, and lorries and the electric grid system make raw materials and power available everywhere, so that industry can be dispersed throughout the markets; now that the government, town councils and boards of directors are making genuine efforts to diversify industry at all levels, and in most parts of the country, the labelling of any one industrial area by means of one industry is meaningless.[14] There may be some point in stressing the original industry, but the disappointing truth in Britain is that nowadays any industry can locate and has located in any of the main industrial areas. Thus while the delineation of industrial regions in Britain has more justification than that of farming regions, the too familiar, too often repeated contrasts in the types of industries are becoming increasingly inaccurate and one may imagine the time when all industrial regions may be described collectively at one and the same time, with perhaps some reference to their markedly different origins during the first industrial revolution.

We have established that the descriptions of small areas alone can be the result of direct individual observation; any geographical description of a whole country must be based in large part on indirect observation, that is through the work of many researchers and by means of maps. Ideally, for a country such as Britain, and certainly for larger countries, maps should be collected or drawn for each topic under consideration, and while it is admitted that these are usually on a scale much larger than that of the maps in *The Atlas of Britain*, this atlas presents ready examples for the reader of the types of maps necessarily involved. The scale of the maps, the method of mapping employed, and the degree of detail involved will lead to many superficial variations, but if the information is correct and plotted accurately, then any such maps of Britain must be fundamentally similar to those in the atlas. If this is agreed, then we are still left with the problem of how the writers of detailed, authoritative regional geographies of Britain arrived at such definite and such closely corresponding regional divisions of the country. If the argument that one can become aware of regional characteristics only by (i) direct observation or (ii) indirect observation, and that Britain is too large for one man to collect all his own information and make all his own maps in one lifetime, is sound, then faced with similar types of maps to those necessarily employed by the regional geographer, we ought to be able to perceive some of the regions he describes so confidently. However, this is certainly not the case, for the maps do not show the boundaries, gradients and vivid contrasts which the descriptions lead us to believe in and expect.

There are several possible explanations for this:

1. As suggested in Chapter 1, the original regional division of Britain, inspired at the time, has been followed uncritically as a matter of convenience since.
2. Geographers employ an *a priori* division based on traditional provinces.
3. The regional boundaries, which the reader is allowed to think mark significant changes in climate, soil, farming, settlement, etc., are in fact based only on relief, or on one other single topic.
4. The regional division is the result of the geographer's analysis of all the topics, and is his personal contribution to make the assimilation of a mass of information so much easier. Being personal, it is not expected to be self-evident. (This possibility would be much easier to believe if many more regional geographers did make their own personal analysis and if the regional divisions thus presented for our consideration showed much more variety.)
5. The regional division is revealed by some method of mapping other than those considered so far.

[1] Kimble, G. H. T., 'The inadequacy of the regional concept', ch. 9 in *London Essays in Geography*, eds. Stamp, L. D., and Wooldridge, S. W., Longmans, 1952

[2] Hartshorne, op. cit.

[3] See Himus, G. W., and Sweeting, G. S., *The Elements of Field Geology*, University Tutorial Press

[4] Houston, J. M., *A Social Geography of Europe*, Duckworth, p. 138, Fig. 24

[5] *The Oxford Advanced Atlas*, J. Bartholomew, OUP, 4th edn., 1931, p. 6

[6] *The Atlas of W. Europe*, Jean Dollfus, Murray, map 25

[7] Houston, op. cit.

[8] e.g. Watson and Sissons (eds.), *The British Isles, A Systematic Geography*, Nelson, 1964

[9] Monkhouse, F. J., and Wilkinson, H. R., *Maps and Diagrams*, Methuen, 1952

[10] e.g. *The Atlas of South East Asia*, Macmillan; *The Atlas of Britain*, OUP; *The Oxford Economic Atlas of Africa*, OUP; Land-Use Survey Maps of Britain (1st series, prewar, and 2nd series, postwar); Ordnance Survey 10 inches to the mile maps

[11] Page numbers refer to 1st ed.

[12] Demangeon, A., *The British Isles*, Heinemann; Ogilvie, A. G., *Great Britain*, CUP; Unstead, J. F., *The British Isles*, ULP; Stamp, L. D., and Beaver, S. H., *The British Isles* (agricultural regions), Longmans

[13] Stamp and Beaver, op. cit., chs. 11 and 12

[14] cf. Freeman T. W., and Rodgers, H. G., *Lancashire, Cheshire and the Isle of Man*, Nelson, 1967

8

THE INFLUENCE OF METHODS OF MAPPING

Once the regional geographer is dependent on indirect observation, then the nature of the media he employs is of vital concern. Not only must his information be complete, accurate and up to date, but it must also be mapped with precision, and by a method which introduces the least distortion. Many types of distortion are so easily possible,[1, 2] but we are concerned here only with the dangers of introducing boundaries where none exists in reality, or of obscuring those which do exist.

Methods of mapping all types of phenomena may be narrowed down to eight in number. The only property of each separate method under consideration here is its ability to reveal significant areal contrasts accurately. Chorochromatic maps, choropleth maps and isopleth maps may be taken as one group, in that these methods of mapping attempt to cover the whole surface and give an impression of continuity in the topic so mapped. The chorochromatic map, which makes use of different colours to show the existence of different things, can be quite accurate in the form of a geological map, but most misleading in the form of a map of natural vegetation. Geological boundaries are real, and different rocks are broadly self-evident, but even so the junction between two strata is often very difficult to pinpoint in the field. Small-scale maps which show sharp boundaries between two types of natural vegetation are, however, most misleading, yet natural vegetation has been taken as the basis of a system of World Natural Regions. These maps tend to overemphasise boundaries and discontinuities. While being drawn in only one colour, the choropleth map presents the same danger, for the steps in the gradation from lightest shading to darkest may again be mistaken for real boundaries. With chorochromatic maps the cartographer intends to show boundaries, but with choropleth maps they are simply a necessary evil inherent in the method. Thus a rainfall map illustrating the range of annual rainfall from the deserts to the jungles of Australia in increasing densities of blue may give the impression that there is some special significance about places with 10, 20, 40, 60 and 80 inches of average annual rainfall simply because the key ascends in these steps. This example is so obvious that no one would be misled by it, but the very same map may well *obscure* a significant increase in rainfall at the foot of the Queensland mountains if that increase takes place at (say) 43·7 inches, somewhere within one of the zones shown as uniform on the map. The most widespread danger of all lies in the atlases used throughout our schools and universities, for without exception layer tints are used to represent relief. The change from one shade of green to the next, or from green to brown to purple, everywhere gives the impression of a break of slope, and even experienced geographers have to keep reminding themselves that the pale green is not necessarily flat, nor the purple always an indication of jagged mountains. Our large-scale topographical maps have progressed from pictorial side elevations of conical hills, through hachures to the precise use of contours and form lines as the only exact method of showing relief, but atlases, which are in much more general use, are still primitive and distorted in this respect. Of the three methods mentioned so far, isopleths, lines joining all places of equal value, are the most accurate and the least misleading. Contours are certainly most reliable to represent relief, and in the case of isobars, isotherms, isohyets and other lines drawn on maps *without* colours or shading in between, one may more easily imagine a gradient between the lines, which is nearer reality than the impression of steps given by the other methods. The more isopleths which are drawn on any map, the more accurate the picture, one has only to compare the contours on sheets of the 2½-inch map and the 6-inch map of the same area to be convinced of this; but even with widely spaced isopleths one is able to interpolate easily, steeper gradients are revealed, significant boundaries are not easily lost altogether, and one line is not artificially emphasised as more important than another.

The other five methods are grouped together here because by nature they give a discontinuous cover. Some topics, such as distribution of population, are most accurately represented in this way, so at the same time both the phenomena represented and the methods of mapping are less likely to reveal obvious tangible boundaries. Individual objects, or groups of objects, may be shown by means of dots, symbols, circles, columns or graphs drawn on the maps. These range in quality from the neat, precise dot distribution map to the clumsy compilation of miniature graphs which obscure parts of the map on which they are imposed. Dot maps are most useful for features of human geography, particularly commodity production and population distribution. They are not likely to suggest boundaries where none exists, in fact their only shortcoming is that if one makes the value of one dot much too high or too low, either they will be so spread out and separate as to give no picture at all, or so close as to be touching and present a black patch to the eye. Symbols may be regarded here as essentially similar to dots in this respect, but divided circles and divided columns may have a different and quite valuable use. Again, if just a few huge divided circles are drawn, as in *The Atlas of Britain*, then their value in the possible definition of regions is almost zero. But where several topics are to be mapped at once, say the distributions of all crops in the country in an attempt to reveal significant crop combinations, then provided as many small circles were drawn as the available statistics permitted, the groupings of circles with large segments for hay and clover in one area, contrasted with groupings of circles with large segments for wheat, barley and roots in another area, might be of use. The idea is virtually the same with divided columns but the important point in the case of either method is for the map to have so many circles or columns that precision and fineness of detail is possible in the demarcation of any region which is suggested by the facts plotted on the map.

Thus it seems that isopleths and dots are likely to be the least misleading. Unfortunately not all the topics usually considered by the regional geographer can be best shown in one of these two ways; structure, soils, land use and manufacturing industries come to mind at once as being more accurately mapped in some other way. So wherever indirect observation is necessary, and one has to rely on maps prepared by others, probably for some other purpose, these questions should be borne in mind: does this method of mapping exaggerate boundaries or create them where none exists, and does this play down possible boundaries or obscure genuine ones completely? The word boundary has probably been overworked here, for in fact usually we find only steeper gradients or discontinuities if we find anything at all. Another point worth emphasis is that the centres of contrasting areas may be much easier to find than their boundaries, and symbols, circles, columns and chorochromatic maps are much more likely to reveal these central characteristics.

Much more important for the study of localities and districts are the maps the geographer may have to make for himself, but by that very fact he should be aware of their shortcomings and know what they can and can not reveal. Haggett[3] has collected together an impressive number of analytical methods which can be of great use to the geographer, a few of them of special interest to the regional geographer who sees the role of geography as defined on p. 14. Given the wealth of detail in Haggett's work, which repays careful study, it is rather disappointing to find that in fact this does not contain a complete solution to our problems. To counterbalance both the optimism which a first acquaintance brings, and the rather exaggerated claims made for this type of analysis, a clear, brief statement needs to be made about exactly what it can do. In Chapter 9 Haggett considers region-building and shows that the methods which he explains at length can:

1. Find precise boundaries for the single-topic functional regions of human phenomena, by:
   Thiessen polygons
   Distance-minimisation functions
   Discriminant analysis
   Graph theory

2. When a number of multiple-topic regions have already been chosen, i.e. countries, states, counties, and parishes with definite administrative boundaries, and larger groupings are required, then they will show which regions are sufficiently similar to be grouped together, whether they are contiguous or at opposite ends of a continent, by such methods as:
   Distance analysis in n-dimensional space
   Chi-squared analysis in classification
   Variance analysis

3. (i) Reveal the different, significant levels of generalisation to show where different types of explanation are essential.

(ii) In single-topic formal regions of human phenomena, reveal trends, slopes, gradients, positive and negative residuals (known by many other names, as listed in the Appendix) in several different ways, and to different degrees of generalisation, but much more accurately than by the rule-of-thumb of the non-mathematical geographer.

Equally important is the need to state some of the things these statistical methods can not do. In 1 above they can not choose the functional or nodal centres for the geographer, nor can they choose his criteria. In 3 above, while some of the maps, diagrams and graphs produced do show quite clearly where different levels of generalisation and different types of explanation are necessary, they do not *ipso facto* reveal or suggest the correct explanations. Much more important, however, and a fact not immediately clear once Haggett goes into detail, is that most of the time these methods apply only to one phenomenon at a time, and the problem of building up the sixteen-topic compage remains. It might be argued that one only has to analyse each of the sixteen topics in turn and superimpose the finished maps or diagrams, but even Haggett himself gives ample evidence, and states that this still does not work even with the refined techniques which he makes clear. His example of resolving the boundary girdle produced by precise analysis in the north-east USA (p. 246) involves the same amount of distortion as the example of resolving the boundary girdle produced by subjective observation in Central Europe (p. 245). However precisely the boundaries of single-topic regions are defined, if they show as little correlation as boundaries produced by more haphazard methods (as they do), then the process of reducing them to a single boundary negates the earlier precision work.

Two interesting points arise in these cases of Central Europe and the north-east USA. The fact that sixteen different geographers produced sixteen different concepts of Central Europe at sixteen different times seems wrong to Haggett; to the present writer this is most laudable when one bears in mind that the human geography of the area has changed so rapidly, and that the sixteen men may have had sixteen different purposes. Secondly, here we have evidence of formal and functional regions overlapping and existing in the same place. Connecticut is seen quite clearly to be part of the formal region of New England, by reason of similar characteristics, and part of New York's functional region as a residential area for the city and in being served by the city.

While the methods of locational analysis will be of increasing use, their advocates such as Haggett make the work *seem* unnecessarily involved, largely because of their present enthusiasm. Thus, when one examines a new technique in his work, part of the disappointment at the end comes from the feeling that one knew all this before, and one has to make an effort to remember that these techniques are not always doing something new—they are doing something geographers have done for years, but now with much greater precision and accuracy. In addition, the scientific method of working common to all these techniques seems unnecessarily tedious in geography. The method is to

1. examine a few examples, very precisely,
2. predict general laws from these few particular examples, and construct a model, and
3. test the laws and/or the model by examining the other examples in real life.

In contrast, in geography, the past tendency has been to

1. examine the majority of cases, and
2. make general laws based on this majority, finding in the process that stage 3 is often pointless as the number of possible permutations in the multiple-topic compages throughout the world is so great that in fact we rarely find two examples of the same kind.

Thirdly, some of these models are more difficult for the layman to understand than the concepts they are supposed to illustrate. The idea of models, and of imagery in literature, is to help explain the unfamiliar by reference to the familiar. The example Haggett gives of Newton's theory of gravity is not as familiar to the layman as the self-evident fact that he and his wife are attracted most strongly to the biggest town with the most shops and entertainments.

In contrast to the enormous help which Haggett's work can give to the regional geographer, and in contrast to the precise thought on actual regional problems, it may do some great damage as the result of the emotional attack in Chapter 1. The impression left by the explanatory chapters and this attack is that all earlier methods are defunct and that the more traditional regional geographers must mend their misguided ways. Certain assumptions not stated in Chapter 1, however, show that the validity of the criticism depends both on the type of geography in which one is interested and on what one considers the main purpose and function of that geography to be. Of the two types, General and Special, and at least three functions, viz.:

1. Areal differentiation
2. The study of the relationship between man and his environment
3. The study of location

Haggett, clearly, is interested in General Geography for the purpose of studying location, and also somewhat contemptuous of the rest. He is concerned with the need to look for pattern and order for the practical purposes of government and business. This is perfectly acceptable, provided it is made clear, but he implies strongly that the interests of others must coincide with his own (and therefore their different methods must be criticised) when he sides with Bunge in stating that 'regional geography [must move] into a search for generic and not unique studies', and with Schaefer in stating that 'regional geography must become the laboratory side of an essentially theoretical subject'.[4] *Must* move . . . *must* become . . . ? Such statements mark the advocate rather than the scholar. Perhaps in the opinion of many, regional geography *should* change, if certain things such as practical application are required; but no one should say what another *must* study. He even presumes to say what regional geographers ought to have been doing, stating categorically that they ought to have been looking for comparisons rather than regional differences. While the attack is alarming, it is not as harsh as it sounds, for in saying what others ought to do, Haggett is simply saying what he likes to do. If written out fully, the argument condensed rather enthusiastically might run along the following lines: some geographers are interested in quantitative theoretical work, and some regional geography can become their testing ground; other geographers have every right to pursue other interests, and other regional work may be qualitative and literary.

By condensing his argument and leaping several stages of logic, Haggett glibly seems to undermine the assertion that regions are unique. He uses Kimble's technique of knocking down one easily defeated tin soldier and implying that this defeats the whole army. No one would pretend that because two pieces of chalk differ in detail it is impossible to classify them as chalk. On this level, the regional geographer is just as ready to agree that stows, sites, tracts, farms, mines, villages and factories are not unique, are not sufficiently different from the rest to require completely separate treatment. Nearer the other end of the scale, however, provinces and realms such as Ireland, New England, Egypt and Java are unique by reason of a unique permutation of all the phenomena within the total compage. We classify people much more readily than some geographers will classify regions, but would the geographer most contemptuous of 'uniqueness' contemplate for a moment the substitution of his wife and children for other equally charming people of exactly the same age, sex, height, weight, colour and nationality? So much depends on personal interest. Some geographers, like biologists, are interested in the type, whether it be the basic type of structure of functional regions, or the basic mechanisms of the human body. Other geographers, like Ed Ricketts,[5] are interested in regional variations of the type, whether these are different compages, or the different shapes of a staggering number of girl-friends.

To conclude the point about how the geographer becomes aware of regions, we see that direct observation of every detail of a continent is impossible; secondly, that even with complete sets of maps available, sufficient regions are not revealed to justify the outlines of regions we see drawn in the textbooks. Therefore we are forced to the conclusion that the geographer must be influenced by other types of information—unless he has admitted openly, and fairly enough, that his regional division is entirely his personal interpretation of the real facts. Of the four possibilities enumerated above, if the regional division is not the geographer's personal contribution, and if we cannot find sufficient evidence for the original division or a clear-cut framework of relief regions, then factors outside the expected scope have been operating. In this case the original ideas for the regional division may well have come from a few scattered regional monographs, from travel books, from fiction,[6] folklore, tradition, hearsay, awareness of local dialects and other traits. We readily accept the fact that a geographer can divide Europe by countries, taking these as his major regions, simply because everyone 'knows' about these countries. Similarly the fact that everyone 'knows', has heard of, the Highlands, Pennines, Lake District, Potteries, Black Country, Weald, Wessex, Connemara, East Anglia and so on, has been regarded as sufficient basis for a regional division along these lines, whether or not these traditional districts have any significant contrasts of any value to the geographer.

Returning to the problem of maps now, having considered the main possible errors introduced by the nature of the maps and the information on which they are based, it seems that the main danger is for maps to suggest the existence of regions much more strongly than the actual evidence on the ground. This is because maps are stylised, simplified versions of the truth and the processes of symbolising the information and reducing it in scale inevitably involve generalisation and division. So in mapping what may in reality be a gradual change from one end of the country to the other, the usual method is to go in steps which at the same time suggest uniform areas on the treads ('regions') and divisions at the risers ('boundaries'). Yet even when the main error in mapping leads towards the suggestion of non-existent regions, even when the map shows more sharp contrasts than exist in reality, we still find that there is not enough evidence for a complete regional division of a country such as Britain which will include every part of the land surface. If we are completely objective about this we are still left with large areas which have no strong regional characteristics and no definable boundaries.

One of the main objections put forward by Hartshorne[7] to the neat regional maps in the textbooks is that when one maps the regional boundaries of each topic in turn, and then attempts to superimpose all these boundaries, they do not coincide at all. The picture he describes is something like that shown in Fig. 5. It is very easy to be mesmerised by his argument, and to imagine each set of single-topic regions mapped separately, and to appreciate the problems of multiple correlation. In contrast, it seems to the present writer that more fundamental problems than this matter of correlation between topics are

1. the problem of drawing boundaries between different regions of one phenomenon and
2. the virtual impossibility of avoiding large gaps.

Thus we are faced, not with fifteen different topics ranging from structure to population, each of which can be divided into regions which do not correlate with the regions of the other fourteen topics; but with topics which vary imperceptibly in fifteen different directions at the same time. This is very difficult to present diagrammatically, let alone to map, and when one faces this real problem rather than the simpler one considered by Hartshorne one can understand why geographers adopt a subjective, artistic approach. Having seen how maps of the various phenomena in Britain vary imperceptibly from one point to another, not always from east to west, but even when the direction is similar then varying in different degrees, one can see that correlation is either impossible—or ridiculously easy.

Hartshorne's argument is that a map of the regional boundaries of each topic can be made on talc, Kodatrace, or celluloid, and then that as we superimpose each successive map on the others, correlation between them gets rapidly worse. We would expect close correlation between the maps of relief and rainfall, rainfall and farming, industry and population, but less with threefold, fourfold combinations and so on. We can appreciate this particular argument so readily, and agree that when all maps are combined there will be

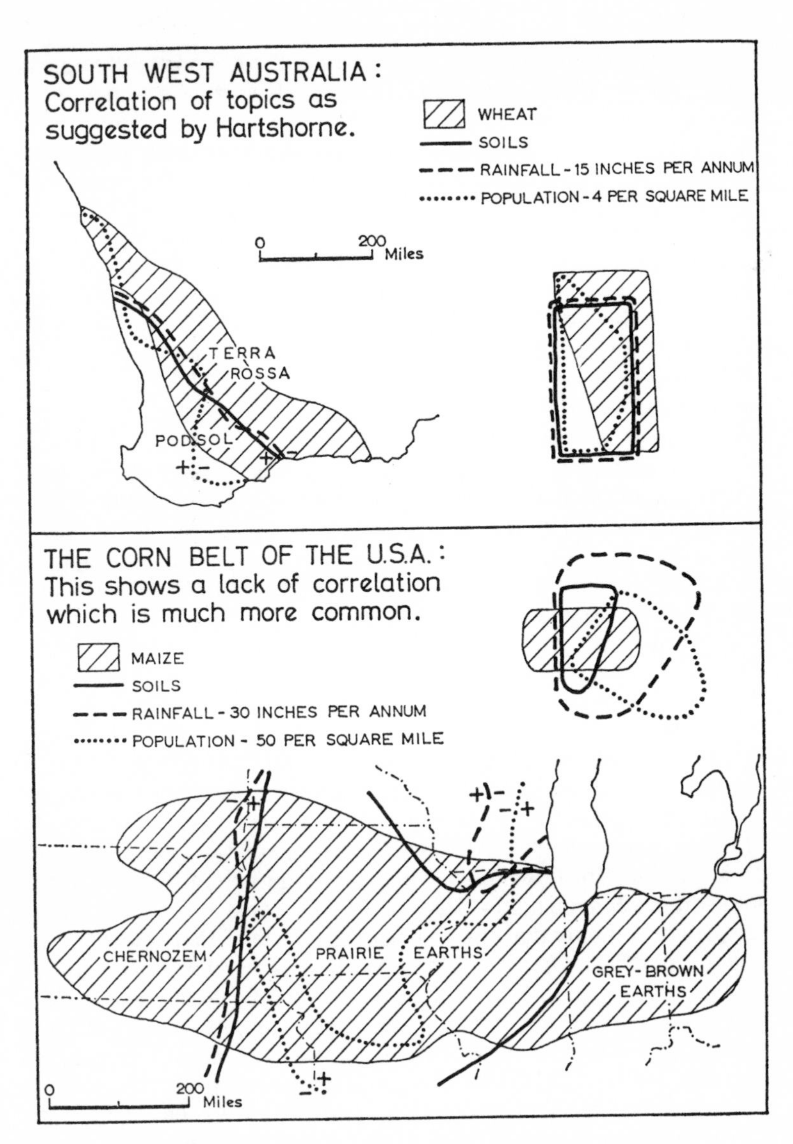

Fig. 5 Correlation of boundaries. The small diagrams idealise the boundary girdle and the lack of correlation

no recognisable pattern left, that we do not question the first step in the argument closely enough. In the last few pages we have seen that this first necessary step in Hartshorne's argument is impossible, but by that very fact the last stage is made infinitely easier. Now the nature of the problem is to superimpose many sheets of celluloid, not with definite lines drawn on them, but changing gradually from being opaque in one place to being transparent in another. Thus they would look more like photographic negatives than overlays for a map. In fact a different colour might be chosen for each topic. When such maps are superimposed the regions of any one topic are defined not by features of that topic itself, but by the way it combines with features of other topics.

This is the fundamental point, and a very primitive example is given here in an attempt to establish it clearly. In the case of Britain, in actual fact rainfall, temperatures, distribution of crops and animals, and similar phenomena do show similar gradations in almost the same direction. The only point to add is the obvious one that as the map of the distribution of dairy cattle becomes lighter eastwards the map of cereals becomes darker. Rocks, relief, mining, industry and population distribution, however, show much more sudden changes, even abrupt edges to observable regions. Thus when all maps are superimposed, providing they are thin enough and the light is bright enough, the sudden contrasts in such topics as relief and the distribution of population will show through those topics which are featureless gradations. These definite areas, then, can serve for a division of all topics. If any one phenomenon changes uniformly right across a country, and one needs to draw a line in order to make smaller areas for easier handling, then the only possible place to draw the line is where one finds a boundary in some other phenomenon which is to be considered at the same time. This is not to imply that there is any causal connection between the two phenomena. Of course, this method is as old as the hills. 'Geographical regions' are in fact usually relief regions, but at the very least it has been implied that there are causal connections between all phenomena in the region. What is needed is a new attitude, greater intellectual honesty, in making it clear that in a particular case relief or population contrasts have been taken as a convenient framework for all the other features which do not show such contrasts but need subdivision for more detailed treatment (see Fig. 6).

This would be the simplest case, but again the actual examples of phenomena in Britain suggest a more sophisticated example where the separate localisations of fruit in Kent, sugar beet in Yorkshire, Lincolnshire and East Anglia, poultry in the Fylde, cotton in Lancashire, whisky in Moray and Banff may each define a unique

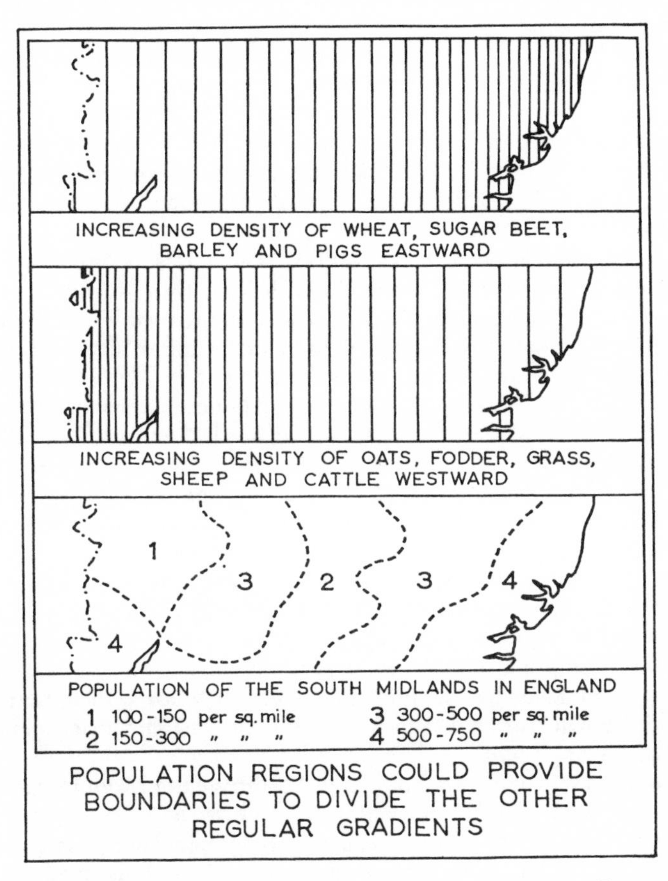

Fig. 6 Boundaries in gradually changing phenomena. For explanation see p. 116

region. The change in criterion in each case is no drawback here. Changing criteria have been observed above in the table of the hierarchy of regions and are quite permissible providing the fact is made clear.

If, with Hartshorne, we insist that multiple-topic regions are built up from the correlation of single-topic regions, then the work is impossible and a satisfactory division will not be achieved. If, with Whittlesey, however, we admit two points

1. that the maps of single topics do not reveal or suggest a regional division
2. that for the final synthesis correlation of all topics neither occurs in reality nor is necessary for our purpose

then all topics of interest to the geographer can be combined, and a satisfactory regional division can be achieved by reference here to a self-evident unit of relief, there to a concentration of population and elsewhere to a characteristic type of farming.

Hartshorne sees the ability to draw boundaries as critical to the whole concept, and if one agrees with him, then we must either abandon the regional method or vastly improve our methods of defining boundaries. But this would be distorting reality even more than is done now. Whatever method might be developed more accurately to define single-topic regional boundaries of the five elements of natural regions, this would not in fact mean that there is any closer connection between them. Hartshorne argues that the correlation of structure, relief, climate, soils and natural vegetation in 'natural regions' is very doubtful, and even more doubtful is any correlation between these and features of human geography. Therefore, he argues, the usual approach through physical features to human features is nonsense. This can be seen quite clearly in the majority of regional textbooks where the facts of the physical environment are given first but very rarely are shown to have much detailed influence on the human activities. It follows, then, that we might be better occupied by taking the facts of human geography as the starting point.

So far in this book, attention has been directed to regions, regional methods, and regional concepts as they exist and as they are in common use. It has become clear that geographical regions are not everywhere self-evident on the face of the earth, and that methods of regional division are not entirely satisfactory. Yet the writer feels strongly that this method of division and presentation is the best yet devised, the one most acceptable to general readers, and the one most likely to give a world-picture. Therefore it is necessary to advocate a method not yet in general use, but which avoids many of the errors of traditional methods. This is the idea of the compage, put forward by Derwent Whittlesey but as yet ignored by many other geographers. The discussion of the compage will contain some additions to the concept—additions of which Whittlesey may not have approved, but the original idea was his.[8] If it should come into general use it might well be adapted out of all recognition, but as will be seen, adaptability to unique, individual cases is one of the key points of the concept. It seems that such a new approach is necessary, not only in the way of defining a region and presenting the facts about it, but also in making clear each time just what criteria were used, just why the material is arranged in this way, and above all that this is the personal, subjective view of the geographer and that it is only one of many other possible points of view.

[1] Monkhouse and Wilkinson, op. cit.

[2] Dickinson, G. C., *Statistical Mapping and the Representation of Statistics*, Arnold, 1965

[3] Haggett, P., *Locational Analysis in Human Geography*, Arnold, 1965

[4] Haggett, op. cit., p. 4

[5] Steinbeck, J., *The Log from the Sea of Cortez*, pt. 1, 'About Ed Ricketts', Heinemann, 1958

[6] Gilbert, E. W., op. cit.

[7] Hartshorne, op. cit., p. 267

[8] Derwent Whittlesey will be regarded as the author; see, however, fn. B, p. 22, in *American Geography: Inventory and Prospect*, eds. P. E. James and C. F. Jones, Syracuse University Press, New York, 1954

9

THE COMPAGE

A region may be regarded as a spontaneous expression of physical and psychological differences.

The regional concept is based on natural or unrestricted relationships between people and places.

In its nature and meaning, a region is of indefinite extent.

These three quotations, taken from the selection given by Odum and Moore,[1] seemed at first to be the most vague, unhelpful, and useless. Yet at this point they contain three ideas which are important in the consideration of the compage. These are, that the region can be spontaneous, unrestricted and indefinite. However the word 'spontaneous' may have been meant originally, it is here taken to mean spontaneity on the part of the geographer in taking a point of view different from his predecessors and fellows. Again 'unrestricted relationships' imply that there should be no restriction in the geographer's mind to observe, describe and comment on any correlation between *any two* sets of phenomena, however out of the ordinary they may be. Finally, the third quotation is the most honest of all, and we may keep nearer to the truth by being concerned with central characteristics rather than by being preoccupied with a pursuit of nebulous boundaries.

Linton put this last point in a different way when he wrote 'The genius of French geographers is for the apt and vivid characterisation of the essence of a region in words'.[2] Linton's unrestricted view of the problem which enabled him to change criteria confidently in mid-classification has been considered in detail above, but he mentions boundaries in the same connection, and states 'Since the basis of characterisation may change from region to region there is logic in avoiding difficulties of drawing boundaries according to criteria that change as the boundary itself is crossed'.[3]

At least four things are attempted in the article 'The Regional Concept and the Regional Method' by Derwent Whittlesey.[4] In forty-five pages are to be found a condensed history of American regional geography; statements of certain aspects of the region not formulated until the time of the article; an exposition of the compage; and some detailed advice on methods of working. It must be said at once that this is a refined and polished piece of work, where every word counts, where nearly every word has a most precise meaning, and where the result of years of careful thought is given in one sentence. In fact one needs to have been thinking along similar lines to grasp the full implications of some remarks, for no point is laboured.

It is vital to realise this in the early pages when Whittlesey is clearing the ground by defining his terms. Having stated that he regards the aim of geography as 'the areal differentiation of the face of the earth', and having first defined the region as 'an uninterrupted area possessing some kind of homogeneity in its core, but lacking clearly defined limits', he goes on to qualify this by adding that the region is 'an area throughout which accordant areal relationships between phenomena exist'. The trouble is the word accordant. If we read it too glibly we are likely to end by agreeing with Kimble that the shortcomings are hidden by meaningless jargon. This word accordant carries the meaning of agreeing, corresponding, connected or even uniform and homogeneous. Yet after very close inspection of this article one is driven to the conclusion that Whittlesey implied neither connection between phenomena, nor complete uniformity. Throughout the rest of the article the definition makes most sense if one constantly bears in mind that accordant phenomena are simply co-extensive and coterminous without any necessary implication that they are in any way causally connected.

Perhaps the only other difficulty in the language of the article is the seeming contradiction between the emphasis that regions are only a means to achieve the aim of geography and the repeated references to regions as if they were concrete objects. This is the same kind of contradiction we make when we write of the sun rising and setting, or moving north in the northern hemisphere summer and south in the winter; it is simply a matter of convenience to avoid rather roundabout phrases each time, but it serves to remind us how easy it is to slide into the use of imprecise language, then imprecise thought, and finally to mistake the name for the object. Whittlesey makes it quite clear that as far as he and his associates are concerned, the region is 'a *device* for selecting and studying areal groupings of the complex phenomena found on the earth'.[5] The region is an intellectual concept created by the selection of features relevant to the interest of the geographer or the problem in hand. Again a single word is vital. The word selection applies not only to the area or areas under consideration, but also to the number and kind of phenomena which will be included. Thus to apply this device one chooses certain topics or phenomena from the whole range of possibilities in order to give the required groupings and regions; and one disregards all topics which one considers to be irrelevant. This has similarities with Hettner's method of choosing regions by and with varying topics, but at this point Whittlesey's device can be mistaken for a blatantly biased attempt to reach certain conclusions by selecting the facts which support a theory and ignoring those which contradict it—the worst sort of regionalism. In fact this could not be further from the truth, but it is a great danger nevertheless.

As the purpose of this chapter is to develop and elaborate some of the points stated briefly, and some time ago, by Whittlesey, the reader who wishes to examine the original statement on the compage is referred to the original article quoted above. From this point on, unless a direct reference is made to a specific point in the original work, some of the ideas, and most of the opinions, are the present writer's additions, based on the outline of Whittlesey's chapter. This outline includes six attributes of the region which merit further attention:

1. Criteria for defining boundaries within each topic
2. Categories of regions, i.e. the number of topics to be studied within each region
3. Characteristics of regions
4. Cores and boundaries of regions
5. Consciousness within regions
6. Compages

1. The alarming statement, suggesting infinite possibility which practising geographers know not to be the case, that 'the face of the earth, with its complex associations of phenomena, could theoretically yield an infinite variety of regional patterns, each brought

forth by the application of different criteria', serves to emphasise at once that the regions are the results of the way geographers look at reality—the boundaries are where they decide to draw the lines. The fact that most geographers tend to draw the lines in the same places, at breaks in slopes, at the 10-inch isohyet, at the 43°F isotherm, at the edge of the forest, between arable and pastoral, industrial and rural, and so on, savagely reduces the possibilities and brings conventional order; but at the expense of stultifying the imagination, conditioning the eye to see the familiar pattern, and encouraging one to use the safe props of ready-made boundaries. While not advocating complete chaos, Whittlesey is at least suggesting the use of original criteria, and that the boundaries of one topic so defined can serve to define the complete region.

These criteria are the exact numerical values which serve to divide the continuous variation of each topic or phenomenon. The example given is of slopes divided from flat land where the gradient exceeds three degrees, but any isopleth may be drawn in continuous phenomena such as rainfall, temperature, crop distribution, population distribution and so on. A most interesting point here is that Whittlesey (and presumably the majority of the contributors to this American symposium) took it for granted that boundaries are *easier* to draw in such a continuity rather than where there is a discontinuity. This point occurs again in his consideration of cores and boundaries where he states that boundaries in continuities, essentially differences of degree, and which are shown by isopleths, isograms or isarithms, are better than any other type of boundary for our purposes. This is because the isopleth of, say, a certain percentage of wheat in a total crop combination is much more realistic than drawing a line and labelling one side the Spring Wheat Belt and the other side the Corn Belt, when there is wheat in both, and neither wheat nor maize is by any means the only crop in either. Moreover, the line is justified by the facts, it is based on an analysis of statistics, and providing the value chosen for the isopleth is significant, so that it separates areas where wheat has different values to the farmers, it carries more weight than any arbitrary line. This point of view is certainly in direct contrast to that examined earlier in the present work when distribution maps such as those in *The Oxford Economic Atlas* and *The Atlas of Britain* were under consideration. There the assumption was made that geographers prefer and look for distinct discontinuities in the phenomena. This assumption is certainly upheld by an examination of most works of regional geography but, as was seen, discontinuities are not found in the actual maps in anything like the number of cases that the regional works would suggest. Thus it seems that Whittlesey's method is much more practical, and much more in accord with the actual nature of things. Rather than waste time hunting for obvious boundaries, being driven in the process to cling to most unsuitable ones, the more successful method seems to be to grapple with these smoothly changing continuous phenomena and impose significant lines wherever they are necessary.

There are several nebulous points which need to be made clear about the choice and application of these criteria. Much of the difficulty in understanding the concept of the compage stems from the fact that we are in the habit of recognising only the single-topic regions of General Geography, or the 'complete' regions of Special Geography; so that usually the treatment in a 'regional geography' book is assumed to deal with all the phenomena in that area. Most of the time, however, whether we bother to remind ourselves of it or not, we are concerned with a mere handful of phenomena, usually more than one, but very rarely more than fifteen or sixteen. Thus the first point to emphasise is that even a complete study of a region contains a geographer's selection of topics. Secondly, when establishing his numerical criteria for the boundaries, he chooses values, guided by his training, experience and observation of the area under study. Another statement of Whittlesey's again suggests bias in this connection, for several times he writes of choosing criteria so that the boundaries of different topics will coincide. He justifies this by saying that an area homogeneous in one phenomenon has value and meaning only when:

(i) it coincides with other homogeneous areas and
(ii) has causal connections with them.

Thus the field is narrowed down first by ignoring, or avoiding, areas which do not coincide, and secondly by rejecting those coincidences which reveal no causal connection. For he is at pains to point out that coincidence of boundaries of single-topic regions does not necessarily mean that one is cause and the other effect. The confusion of coincidence with correlation is much too common in regional texts and in practical fieldwork, especially in the very facile 'explanations' of the location of industry. Again it is emphasised that Whittlesey's chosen word 'accordance' refers to the co-extensive and coterminous nature of any two or more single-topic regions without any implication of interconnection or interdependence. One must search for coincidence—and then reject those examples which are coincidence and nothing else.

If one is choosing criteria to lead to coincidence, there is the danger of *a priori* reasoning and of prejudice. Therefore the criteria *must* be adapted as the study proceeds to give the closest degree of significant accordance. These numerical values must be revised until they are so accurate that a statement such as 'x% of arable land out of the total farming area is found on slopes of less than y° where rainfall is between z and n inches, there are at least p months above q°F and soils are of types e, f and g' is everywhere true within the limits of the chosen generalisation. So while one is aiming to make correlations, in the end these are based on facts observed on the ground and on the maps; the criteria are adjusted to fit the facts, not the facts to fit the boundaries. Moreover, a statement such as that about arable land above applies *only* to that region. A change in any one phenomenon in any other region renders the whole complex false, and criteria must be re-established to fit the new region accurately.

Finally, the region is homogeneous *only* in terms of the applied criteria. Again we are familiar with climatic regions, and accept that a cool temperate western margin region is not necessarily homogeneous in its relief, structure, farming, industry or population distribution, but only in respect of temperature, rainfall, pressures etc. defined by the criteria of certain degrees, inches and millibars. What we tend to fail to accept is that a 'complete' region is homogeneous only in terms of

Relief defined by the criteria of height and gradient
Climate defined by the criteria of temperature, rainfall, pressure
Soil defined by the criterion of soil profiles
Vegetation defined by the criterion of plant species
Land use defined by the criterion of production per unit area
Population defined by the criterion of numbers per unit area

and not necessarily in any other feature or topic such as language, religion, politics, house types, dress, or diet. One of the worst disgraces in our school textbooks is the statement or implication that natural regions, defined by climate, soil and vegetation, are also homogeneous in structure, relief and even 'Man's reaction to environment'. Similarly, even in advanced and respected texts, the homogeneity in physical phenomena and economic activities is allowed to imply homogeneity in national characteristics, dress, diet, race, aspirations and so on, as though each region were in perfect political and social accord. However, this stems as much from the lack of moral courage to tackle such controversial topics as from anything else.

2. The categories and characteristics listed by Whittlesey are fairly familiar, and have received some consideration (see pp. 27–46). However, one or two points need amplification. The three main categories are single-topic regions, multiple-topic regions and total regions. Multiple-topic regions are divided by the closeness of the connection between the various topics. Thus there is an intimate connection between the topics of pressure, rainfall and temperature in a climatic region; less connection in, say, land-use regions; and least of all in what are called 'associations', such as the traditional natural regions. Associations do not necessarily involve a greater number of topics than, say, those of a climatic region. It seems that multiple-topic regions have at least two topics, and in practice not more than about ten. Total regions, moreover, certainly do not involve all possible topics found in one place. This is where the practical Whittlesey differs most from the more academic Hartshorne. Whittlesey limits the topics, even in a total region, or compage, to something between fifteen and twenty. The field is narrowed down by concentrating on 'human occupance of area' and selecting those features 'believed to be relevant to geographic study'. This is another vital matter which needs to have attention drawn to it. At one and the same time teaching, familiarity with regional texts, habit and common practice all mean that we know which features are usually relevant to geographic study, and that we maintain these familiar habits and continue to plough the same old furrows. Common practice saves us from the horror of a blank piece of paper and infinite possibility, but should serve as a starting point rather than an immutable set of rules. Thus the present writer would emphasise that the geographer may select any topics which *he* believes to be relevant to his particular study of a particular area, and he will be guided by his experience, his particular interest, and the purpose of his study; but the selection from total chaotic reality is his.

3. To complete the categories, Whittlesey includes the possibility of any type of region being formal or functional, which he calls uniform and nodal respectively. Then the following characteristics are brought together:

(i) All regions are

(a) unique (if only in position),

(b) three dimensional,

(c) coherent and accordant (in Whittlesey's scheme of things regions are accordant by definition, for unless at least one single-topic region accords with one other single-topic region, no region exists),

(d) changing, i.e. the present is just one stage in the historical development,

(e) part of the hierarchy (see below), and

(f) defined by their own criteria.

Here it seems that Whittlesey closes the door firmly against a very useful development of his ideas. His point is easy to understand and accept when, for example, he insists that climatic regions must be defined by the elements of climate, and not by elements of natural vegetation, soil or even relief. Geographers have been embarrassed too long by the climatic regions of Köppen and Supan which have proved on examination to be based on natural vegetation, which no longer exists in many parts of the world. However, the nice distinction could be made, that while a certain uniformity in climate does not imply a perfect correlation with wild vegetation and zonal soils, a uniform region defined by its climate could be made the *device* for a concurrent study of the vegetation and soils, provided that it was at all times made clear that the regions of uniform vegetation or soil might be larger or smaller than the established climatic region. The necessity of constant vigilance on this point would be worth the great convenience of this device, and involves greater precision of thought.

Here is the chance to face up to reality and to reduce our problems at the same time. As Hartshorne showed, the boundaries of fifteen single-topic regions in the same place do not coincide, yet there is an interest and a need to consider these fifteen topics in the 'complete', total region or compage. While Whittlesey did not go the whole way, his attitude to the problem, that of re-defining and re-drawing the boundaries, might lead some workers to make all boundaries coincide. However, it seems better to agree with Hartshorne that in fact both self-evident boundaries (discontinuities) and imaginary boundaries (based on significant criteria such as 43°F) do not coincide, and rather than beg this question, as the majority of authors of regional texts have done so far, state clearly that one's region is defined by the boundary of one topic, however many other topics are to be considered.[6] This would be perfectly in accord with another of Whittlesey's points about the compage, that *any* of the fifteen or so topics may be the starting point, the theme, the *raison d'être* for the regional treatment.[7] Thus a theme for a part of Yorkshire might be the industrial region with its great concentration of wool textiles, and the region defined by the distribution of woollen and associated mills. In contrast the theme in parts of Lincolnshire, Cambridgeshire and Norfolk might be the distinct features of the economic and social geography, with the emphasis on specialist farming, with the region defined by the extent of land reclaimed from the fens.

(ii) Returning to the brief list of regional characteristics, let us note that only formal, uniform regions are

(a) homogeneous in the phenomena chosen, within the limits of the criteria, which still implies much generalisation;

(b) alternatively, homogeneous in the sense of having a recurring, often repeated pattern; i.e. any one farm has different crops in each field, but all the surrounding farms show the same combination of crops.

(a) might be compared with a Lovat Harris tweed which is in fact made up of blue, green, brown and even red threads but which the eye generalises to one colour, while (b) might be compared to a Paisley pattern.

(iii) Only functional, nodal regions have

(a) homogeneity in the sense that all the contrasting parts under consideration are focused on a central place;

(b) a central place, focus or node which may be connected to the region in one of several ways. It must be remembered constantly that this is the focus only of those topics under consideration, but the centre may be the focus of two regions in different ranks (sizes) of the same category (number of topics), or even in different categories. Moreover, the focus of an economic region may actually be outside the territory, as Oporto is outside the area of vine-growing for port wine, or Rotterdam is outside Switzerland for which it acts as the main port;

(c) a pattern of circulation, including lines of flow such as roads, railways and canals dealing with material transport, and/or lines of force such as political and religious ties with the central place;

(d) declining intensity away from the centre and in interstitial areas off the main lines of circulation.

While some homogeneous formal regions decline towards their edges, within the upper and lower limits of the criteria, e.g. a rainfall region declining in one direction from 20 to 10 inches, this decline in all directions is an essential of the functional or nodal region, as the boundaries are drawn where the pull of two centres is equal and opposite, and by implication, rather faint. The declining intensity off the main lines of circulation is not so generally recognised, nor so well emphasised. The actual surface of a functional region is more like a very complicated spider's web, with the similar gaps and holes, than it is like a set of concentric contiguous rings. Christaller pointed this out in his hierarchy of towns, where insignificant villages exist between the metropolis and the first ring of large regional centres. McCarty remarked that the banks of the St Lawrence estuary, one of the main 'lines of flow' between America and Europe, are among the most remote places on earth, and certainly the Essex marshes or Formby sands can feel the loneliest places in the world.

4. In connection with this characteristic of decline from the centre are the attributes of cores and boundaries. Obviously, in the functional region the core is that part of the region most tightly connected with the node or focus, and is usually the central area. However, as generalisation is essential in geographical description, some parts of a homogeneous formal region will only just have the essential criteria, while other parts will in fact be much nearer the optimum conditions. These latter parts form the core of the region, where the chosen regional characteristics are clearest and purest. In the case of formal regions this core must be different in character from surrounding cores. In contrast, as discussed at length much earlier in this work, several similar functional regions may exist side by side, and be properly described at one rank in the hierarchy before being grouped together as one formal region of the next higher rank. Whittlesey wanted the emphasis to be put on cores rather than boundaries, because with our constant preoccupation with boundaries we often lose sight of their purpose—to separate *different* regions. Whatever leads geographers to regional division in the first place, it is certainly not the existence of boundaries, or any burning interest in them. One arrives at regional division by interest in different parts of the earth's surface, by a realisation that one core is different from another core, and then one is forced to try to establish some sort of boundary between them. But the emphasis and the interest should remain with the characteristic centres of the compages, the cores.

Thus the things we tend to mark on our maps are the least important things. We mark the boundaries, where there is nothing vivid, rather than the centres which first attracted our attention because of some interesting feature. Perhaps one reason is that while the correct place for a boundary line is very hard to find, at least we know how to mark it when we have found it—with a line; but we have no method for marking the main characteristic of the centre of each region. In theory every possible region on the face of the earth would require its own unique symbol.

For the moment, however, regions are shown on maps by one or more of the following types of boundaries:

(i) In discontinuities—difference of kind often made by Man or emphasised by him, e.g. political boundaries.

(a) Lines—e.g. where desert gives way to oasis,
junction of limestone and grit,
the wall between the last field and the first house of the town.

(b) Zones (α) where one region grades into another, e.g. coniferous forest into tundra,
arable into pastoral.
(β) where one is expanding into another, e.g. the city growing out into the county.
(γ) where there is a periodic fluctuation, e.g. a given isohyet is in a different place each year, farming regions wax and wane in response to demand.

(ii) In continuities—difference of degree. Marked only on maps, by imaginary lines, contours or isopleths, and as stated earlier, regarded by Whittlesey as much more common, useful, and easy to define than type (i). A line may be drawn for a chosen value of height, slope, rainfall, temperature, production or population density, etc., even if the gradient is the same on each side of the line.

In both cases Whittlesey mentions 'boundary girdles' where, although the boundaries of several topics do not coincide perfectly, they are sufficiently close to reveal a zone where a refinement of the criteria may reveal significant correlation. But, like Hartshorne, Whittlesey is too optimistic about the universal occurrence of boundary girdles.

5. One of Whittlesey's final comments about this attribute of regional consciousness is that people are aware of belonging only to compages, if they have any regional consciousness at all, and not to single-topic or multiple-topic regions of lower category. It seems important to state the obvious that regional consciousness here means an awareness among a certain group of people that they are the inhabitants of a unique region, as this point may in fact decide

whether regions actually exist or are mere abstract concepts. This is the essential difference between perception on the part of the observer, and self-consciousness on the part of the observed. The observer with his bird's-eye view, the geographer, has the ideas of regions in his mind. These he communicates to his disciples and readers, and unfortunately for any hope of scientific detachment, many of these readers are inhabitants of the areas in question. While some people may have been aware that they belonged to the Highlands, the West Country or the Weald before geographers studied, marked off and labelled them, this is not so sure about people who are now aware of living in the West Midland Conurbation, Merseyside or a Stockbroker belt. The communication from analyst to the analysed is credible enough, if only at second hand when the newspapers and television pass on potted and predigested geography and sociology to the masses. But communication of the essential data about regional consciousness from the inhabitants to the geographer is not credible enough.

The fact that some areas have long-established names is not proof enough of regional consciousness on two grounds. (i) The name may have been given by one man in the remote past. (ii) Literate, educated, informed people may be aware of the name, but not necessarily the majority of the ordinary people. For regional consciousness to have any standing as a criterion for the delimitation of a region, or as a characteristic of many, if not all, compages, then the majority of the people must be aware of the unique area in which they live. As the quantitative data are not normally found in census returns, Board of Trade publications, or the statistics of the Ministry of Agriculture, Fisheries and Food, it would seem that regional consciousness can be a serious factor only if the geographer has the completed questionnaires from everybody inside and beyond all his regions, where the following questions have been posed.

(1) What is the name of the region in which you live?
(2) What is special about it?
(3) How do you know the name?
(4) What other regions surround you?
(5) Why is your region important?

These questions are worded to assume two things which are not necessarily true; that the person answering does live in a region with a name, and that it is important. This is to be as direct as possible, because answers to number 1 such as 'I don't know' and 'What do you mean?' are then most significant. Number 5 will directly reveal the fallacy of special pleading if regional consciousness does exist, especially as much of this is connected with local feeling of the need for special attention or treatment. If the answers to numbers 1 and 2 are similar, e.g.

Q. (1) Answer: 'Yorkshire'.
Q. (2) Answer: 'Well, it's Yorkshire!'
or
Q. (1) Answer: 'Cheshire—Prestbury, actually.'
Q. (2) Answer: 'My dear fellow!'

then one may be confident that there is a certain amount of regional pride, or self-satisfaction. The answers to questions (3) and (4) should be equally revealing, if the person is quite honest, or can remember. Many people seemed to be completely hazy about where they lived until little vans with North West Electricity, Southern Gas, or British Railways Midland Region painted on them started to run round the streets. Moreover, if anyone can rattle off the names of surrounding regions or the names of neighbouring counties, then eliminate his questionnaire; he has been reading geography books, or looking at an atlas, and has been preconditioned to give the answers you want.[16]

The annoying point is that regional consciousness patently does exist, yet it is practically impossible to employ it as a definition or important characteristic of a region. In its better manifestations it can be constructive, in the sense of people working to improve their region, develop its resources, or enlarge its territory. In the southern continents which are obviously undeveloped, or in the development of Benelux or the Common Market, where rapid changes were not expected, this may at least suggest a theme for regional treatment. On the other side of the balance are the destructive manifestations of regional consciousness, varying in vehemence from 'Home Rule for Scotland' through 'Hands off Wales' to 'We can smash England. Join the IRA'[8] and including nationalism which implies antagonism against other nations, and sectionalism which implies antagonism against central government. These manifestations are even less easy to handle, but it is the author's contention that if a geographer is to consider regional consciousness, then he must deal as fully with regional unity against the outsiders as much as with regional unity for some internal purpose.

For regional consciousness to have any value, other than as a possible clue to some local characteristic, at least 51% of the people in any sample must have been aware of their unique region from an early age. If this were so, regional consciousness could be a vitally important basic topic, with the criterion of a majority agreement on the unique character. The boundary of each region would be drawn where majority agreement ceased to be found, and according to the scale of the study and the maps, could be plotted as accurately as any population statistics. The main disadvantage is the difficulty of collecting the facts, which would be feasible only if relevant questions were included on the forms for census returns; and after the first such census everyone would be self-conscious about this matter and distortion might follow later. Without reference to any specific samples he has made, Whittlesey implies that regions defined by 'regional consciousness' do not coincide with what we normally think of as geographic regions (i.e. defined by the geographer) but again he does not say whether he means formal or functional regions. Sectionalism in the United States of America carries the idea that the people in a certain area, aware of real or imagined local problems, organise themselves in one functional unit or formal area which does not coincide with the preconceived ideas of some geographer or government official. Because this reality conflicts with the assumptions of the expert or the official, it is too readily frowned on. We have seen earlier, however, how self-conscious city-regions are one type of functional region which cuts across formal regions, and how this regional consciousness may be manipulated by politicians and professional planners for their own ends. Thus regional consciousness may be more a political than a territorial feeling, and lead us on to the dangerous ground of regionalism, or of geopolitics. On the other hand, while for practical and political reasons regional consciousness may not be universally applicable, in some few parts of the world it may be so strong, clear cut, easily defined and important (e.g. in the Basque country) that it may form the basis and theme of a compage.

6. The difference between region and compage is the difference between theory and practice. Whittlesey pointed out that, until that time, the unqualified word 'region' had implied the totality of phenomena within a given area. Although very little of the theory concerning regions had been written down before his essay, it seemed fair to assume that geographers conceived of three basic categories, the single-topic, multiple-topic and total-topic regions; these latter being simply 'regions'. Yet in practice, one may establish, by diligent examination of all regional texts, that while very few geographers have consciously limited their studies to a given number

of topics, the number usually falls between fifteen and twenty. Thus those who suggest that a regional study is beyond the powers of one man may well have this nebulous total-topic region in mind, while successful practising geographers have in fact been concerned with what Whittlesey named the compage. Realising that the word 'region' is so widely used inside and outside geography that it is a waste of time to try to give it more precise meaning, he revived the old English word to apply strictly to the region of highest category which deals with a definite collection of topics.

This is where Whittlesey stops, and where the present work suggests what seem to be fair developments of this idea. First, Whittlesey was very anxious to avoid a rigid number or list of topics which could become dogma. The present writer feels that such a list is a useful starting point. Second, it can be emphasised much more strongly that the concept of the compage differs from the concept of the region not only in the number of topics, but also in the selection and arrangement of those topics, and in the definition of the areas in question.

The crucial question now arises, which topics *do* we select from all the phenomena on the earth's surface; where do we start in our geographical study of complex, chaotic reality? Whittlesey's answer is one of those annoying statements which succinctly sum up what the expert knows, but which convey precious little information to the novice. 'The compage is, by definition, something less than spatial totality; but it does include all of the features of the physical, biotic and societal environments that are functionally associated with man's occupance of the earth.'[9] Thus one must select those physical and human phenomena connected with man's occupation and use of certain areas of the earth's surface. Here one can do no better than repeat Whittlesey's warning that it is dangerous to start with a standard list of phenomena, and a standard list of criteria. The most accurate characterisation of a compage may involve fewer or more phenomena than the standard list, and may necessitate unique criteria for that unique compage. Yet the real danger lies not so much in the standard list, as in rigid, mechanical adherence to it. Therefore the following list is provided as a starting point only. The writer has been impressed frequently by artists who seem to know exactly where to put the first line on the paper, or the first brush stroke on the canvas. There is no secret about this—they don't. But once they have some line or colour, they have a start, and can build on this. Often the first marks, so vital on the virgin sheet, are of minor importance in the finished picture, or are obliterated entirely. The following list is as dangerous or as useful as a gun; depending on how it is used it can earn the user's living for him, or kill him.

1. *Structure.* This may be important in the compage because of its effect on relief, drainage and soils, and its exploitation by man for minerals and water.

2. *Relief and drainage.* This is an essential feature but the way it is treated will depend on the compage.

3. *Climate.* This major factor in farming is one over which man as yet has little control.

4. *Soils.* In contrast soils do not necessarily determine man's activities for man can change them considerably by marling, plowing, irrigating, draining and fertilising.

5. *Vegetation.* Wild vegetation may be of paramount importance in the Amazon Basin or New Zealand, but of no importance at all in the settled parts of Europe. This is a perfect example of a topic which may be the characteristic of one compage, and which may be ignored completely in another.

6. *Wild fauna.* Again, in some compages this may be ignored, while in others caribou may be an important food supply, rabbits and red deer may threaten the pastures, mosquitoes threaten human life. However, while vegetation tends to receive unnecessary attention, the fauna is often ignored, even when animals are important to the economy and insects affect settlement.

The topics above include all possible physical phenomena to be encountered in a compage. The order may be changed, some may be ignored, and, equally important, the length of treatment of each topic may be varied enormously according to the nature of the compage. A regional geography is not an encyclopaedia. In practice we find that the next group of topics include Man's activities which (a) necessarily involve the use of specific parts of the earth's surface, and (b) usually result in material, directly observable modifications of that surface.

7. *Gathering food* and 8, *Hunting/trapping* may require serious consideration in a few parts of the world but in many are completely redundant. This list includes everything which might need consideration, but many topics should be ignored in a specific compage.

9. *Fishing*, 10. *Mining*, 11. *Forestry*. Topics (7) and (8) may be missed from a compage because they have died out, while these three obviously depend on the existence or not of the resources.

12. *Farming.* A general word to cover the great variety of activities,

one or more of which is likely to be of major importance in the compage simply because of the area covered and the close connection with physical phenomena.

13. *Manufacturing* or *industry*. This is a familiar major topic in regional geography. Many descriptions of industrial regions concentrate on the scattered factories and towns, ignoring hundreds of square miles of intervening rural and residential areas. This is the compage approach in practice if not in name.

14. *Transport* and 15. *Settlement* are obvious enough in the landscape to be included in the compage, but here the *lack* of adequate transport and communications may form the main theme, as in certain parts of Africa and Latin America.

16. *Population*. This is a key topic but its position in the description is not fixed. Often population is treated towards the end where the density and distribution are seen as the results of physical and economic factors. They may equally well form the starting point leading to a study of all the relevant factors, including social factors.

Although this list is based on the content of actual regional descriptions, the fact that most of them include no more than these sixteen does not mean that these are the only topics which could be included. It seems that adherence to a sixteen-topic compage is simply habit, possibly arising from the care to include every agreed topic to avoid censure for omissions, and to exclude any original topic in order to avoid ridicule. However, some guidance in the choice of other topics is necessary, and the main possibilities seem to be the study of:

(i) all material features of the landscape,
(ii) all methods of making a living; and
(iii) other aspects of social life.

Pursuing (i) too far might just be pedantic but such things as field patterns are interesting and significant. In the case of (ii) the only methods usually considered are those which involve definite structures such as farms and factories, and have definite locations or localisation. Methods of making a living such as shopkeeping, medicine, hairdressing and acting are not considered because geographers seem to assume that all these services are adequately provided in every town. Like other topics, the main characteristic may be the relative abundance or lack of services in, say, London and Fort William, rather than the predominance of a single service.

In the case of (iii) the study of ethnology, or the habitat, economy and society of certain peoples, gives us a guide. Interpreting the word society widely, the following topics may be considered.

17. *Religions*, 18. *Politics*, 19. *Social organisation* and 20. *Social life* to the extent that they influence economic activity and lead to a characteristically regional, unique way of life, possibly with a unique cultural landscape. The first three definitely are major factors in the economy, while the provisions for social activities may produce halls, sports grounds, national parks and so on. Moreover a careful comparison of the weekly routine in two different areas, say Norway and Java, reveals fundamentally different characteristics which may be the essence of the compage. The essential nature of the compage is that the inclusion or exclusion of a topic will depend on what is actually found on the ground, and what the individual geographer finds interesting and important, so this list is not closed to other possibilities.

Another extremely valuable result of the idea of the compage is that not only does it force one to re-examine the idea of the region, but also that it forces one to realise that research and writing are two completely separate activities. Far too much regional geography is commonplace and pedestrian because the familiar fifteen topics are treated in exhausting detail, regardless of relevance, and it is perfectly clear that the facts of each topic have been collected together and written down in automatic fashion. Either there is little connection between the topics, or connection is achieved by repetition, for example by referring to facts of relief, climate and soil a second time, in the section on farming. This boring result can be avoided by separating research and writing by a suitable period of digestion and meditation. As one does not know what to put in, leave out, rearrange, interconnect, emphasise or play down until one has the facts, the research obviously can be automatic; working through topics 1 to 20 conscientiously in that order.[10] But they do not have to be presented in the same order, in the same detail.[11]

Most of the following section is a paraphrase of Whittlesey's advice, but with the present author's preface—*Once all the facts have been assembled*, select those items relevant to the compage and present them in order of significance. One has to know the compage, probably indirectly, before one can do this, therefore the final description cannot be written until all the facts are in and digested. Then, if some aspects of the physical environment are seen to have little effect on the total regional character, they can be played down. In contrast, the structure of the Jura, the relief of Norfolk, the vegetation of Brazil, the fishing of Iceland, farming of New Zealand or population of Indonesia might form the central theme to which all other topics were related. The compage *is* a subjective device.

While we are prepared to concede that single-topic and multiple-topic regions are artificial, the case has not been so clear with the 'total' region until Whittlesey revealed it for the sham it is—a compaction of a mere 15–20 topics out of total reality; a compaction which he called the compage. As the choice of a certain number of topics at once makes the device subjective, there is no point in trying to be objective from that stage onwards, but much to be gained by a personal rearrangement of the topics, emphasis, and theme. Again we have to face up to the fact that geography is not a science. If we treat it as a science we lose everything; if we treat it as one of the humanities and put the human point of view, geography could be more interesting and of more value than history. Comparison with natural science and history is pertinent here. In those disciplines there is always the twofold value of intrinsic interest and practical use, although their relative values vary. Nowadays the emphasis is placed on the applications of natural science, its possibilities of profit for industry and government, and material contribution to society. In contrast, the cynics say that the only thing we learn from history is that nobody ever learns from history, and the same mistakes are made over and over again. Yet a knowledge and sense of history can be of extreme practical importance in public life, especially in politics.

Modern geography, however, is not outstanding, either as a fascinating study such as history, or as useful knowledge such as a natural science. One or two geographers stand out for their attempts to show how geography can be useful, notably L. Dudley Stamp, in many books since he was first concerned with the land-use survey,[12] and T. W. Freeman and J. N. Jackson more recently.[13, 14] To make regional geography absorbing and useful, paradoxically, it must strengthen its ties with history. The emphasis should be placed on human geography, on what real people are doing in other parts of the world; if history shows a bias to politics, and plays down the economic and social aspects, geography certainly has a bias to economics and plays down the political and social—Man's social condition, in the long run the most interesting and the most important, is largely ignored. Some geographers go forth to show other countries how to grow groundnuts and kill mosquitoes, but for our arrogance we might learn something much more fundamental in return than the secrets of the wheelbarrow or curing tobacco; we might learn how to live.

One other practical problem faces the geographer. It is all very well knowing what topics to consider, and what type of information has to be collected, but how does one know where the compage is, in order to start to collect the facts? Some compages, a very few, are self-evident. Others become easier to see with experience, and this intuitive, subjective approach is part of the nature of geography.

		Maps	
Name	*Salient points*	*for study*	*for presentation*
1. *Locality*	A small functional region. Village market area or parish. Functional neighbourhood in town. Small enough to give the detail of a farm, or of buildings in town.	2½″ 6″ 25″ to the mile	1″
2. *District*	A group of similar functional units. Homogeneous by repetition of components, e.g. Lake District, East Anglia, Beauce.	1″	¼″
3. *Province*	Fenneman's province (earlier table) e.g. Pennines, Paris Basin. A large region, mainly functional, but too large and varied for every place to be part of one functional whole. Such internal variety that every statement is a generalisation. Direct observation gives way to samples.	1:1M	1:5M
4. *Realm*	Collection of similar or merely adjacent provinces. This is the scale often used for teaching in schools and general texts. Generalisations are so great that many important features are left out.	1:5M	Smaller than 1:5M

cf. table on p. 50

One recent work of regional geography is rare in actually making this clear. In his introduction to *The Western Mediterranean World* Houston[15] states 'The regional landscapes are then finally described, with a bias to their subjective impressions, in order to appreciate their atmosphere, composition and unique character.' However, one very often has to start in the dark, start somewhere, anywhere, in order not to die of starvation exactly halfway between two equal piles of carrots because of indecision. There are many suggestions earlier in this book of suitable units, especially using relief features and administrative or political boundaries as a guide. Then, just as full knowledge of the fifteen topics suggests what to leave out and what to emphasise, knowledge of an arbitrary starting area will give some guide as to whether the compage is bigger or smaller than that area. One must think outside the problem. One must know something about the land outside the original starting area in order to see whether it is different from or the same as the land inside. If it is different, well and good; if it is the same, one must extend the area of research and make a bigger compage. However, there may also be so much contrast within the starting area that subdivision is necessary and one may end up with one or more smaller compages. The only definite units are the continent and the site.

Whittlesey suggested a hierarchy of compages based on the amount of generalisation, and this is set out on page 139 as a table, with certain additions which attempt to make it clearer.

Each of these ranks of the hierarchy may apply to any category of region, as is shown in the simple table below, which serves the double purpose of showing how regions and the compage are related and of making the ideas of rank and category more familiar. Familiarity with, and precision in, their use could make reference to regional geography much quicker and clearer. For example, any book could be labelled by rank and category; or in an examination syllabus it could be made quite clear exactly at which rank each of the categories is to be studied—in the present G.C.E. ordinary examination in England, Britain is usually studied by districts, Europe by provinces and the rest of the world by realms. In fact a set of world maps has been suggested, showing the division of climatic, vegetation, agricultural and industrial facts into realms to provide a basic framework for subdivision into provinces, districts and localities. However, there seems to be no sound reason why a geographer who can define his own provinces, districts and localities should not define his own realms.

Rank	*Category*			
Realm	1 Single-topic	2 Multiple-topic	3 Total region	4 Compage
Province District Locality	e.g. relief	e.g. climate farming	15 topics in rigid order	varying number of topics in the necessary order

As the word 'region' will continue to apply to any of these sixteen possibilities the difference between columns three and four is simply one of final presentation. The total region is the old, too familiar region which deals with all the topics usually regarded as proper to geography in systematic order, often giving the impression that this is being done just to work through one region after another. The compage may vary the number of topics, certainly varies the order and emphasis as appropriate to the region, and has a theme which at least suggests some real interest in, and original approach to, each region.

Some central theme is essential to the compage, and this brings into question both the methods and the purposes of this work. Ideally, the theme comes out of the region, and is in fact the expression of its most important characteristic for the geographer. Being realistic, however, one must admit that some theme can be imposed on an otherwise pedestrian description, or a theme conjured up where none has been suggested even after extensive study. The spirit of the work may be ruined by the same urge to systematic completeness which makes so much regional geography irrelevant to the region in question. Just as was suggested much earlier, the whole surface of a country or continent is always covered by a regional division, whether this is justified or not, there may be an urge to find a compage theme for every part of the land surface, and the work becomes merely 'compage hunting'. To keep in the right spirit, then, the best thing is to make certain topics the foundations for everything else. Dryer pointed out that just as vegetation is the resultant of climate, soil and relief, and can bind these topics together, so the economy is the resultant of human action on the physical environment, and can bind together human motives and functions with the facts of the physical environment. Incidentally, such an approach safeguards against obvious determinism. Climate may be made the central topic in the consideration of physical geography, being influenced to some extent by relief, which in turn is influenced to some extent by structure. Climate then affects vegetation and soil. The whole point here is that for the purpose of the compage description, *all* we need to know of the structure is where it has some effect on some other feature; of the relief, where it has a direct bearing on microclimate, soil, farming, settlement or communications, and so on.

By taking the economy, or land use, as the resultant of the interaction of social and physical forces, the emphasis is placed on the cultural region. Here cultural is used in the wider sense of including the material culture, and the compage is defined by some criteria of the economy in contrast to the usual definition of the 'total' region by the criteria of relief and drainage. At once regional geography becomes much more difficult, because cultural regions change as societies and their economies change—the good old relief regions were safe and static. This gives the challenge of fluidity of place and time. As the society and economy changes, each geographer can redefine the compage according to his own clearly stated criteria, and according to the changing facts. The need constantly to rewrite compage geography will give the opportunity for each geographer to put forward his own point of view, to give his interpretation of the facts; but all this, of course, calls for geographers who have real concern for a compage. Three advantages of this challenging and difficult approach and method of presentation put the value back into regional work. Compage geography compares with history in its variety of bias and interpretation, and in its genuinely educational value. Secondly, it puts the emphasis back on human cores rather than physical frameworks; and, thirdly, it forces the geographer to make his own decisions of time, place, criteria, and the number and arrangement of his topics.

Earlier, it was argued that the methods of working up or outwards from one point, and of dividing or working downwards are complementary, and not mutually exclusive in regional work. However, there are some aspects of these two methods which are likely to affect the theme of the compage, and thus affect the final presentation of the facts to the reader. The inescapable point is that the method of working outwards is suitable only for one isolated region, for one compage at a time. Choosing many arbitrary centres and working outwards simultaneously, or one after another, is tedious and unnecessary. Thus this type of work is characterised by interest in one special region, and by either the complete lack of any standards of comparison or, what may even be worse, unstated, prejudiced assumptions of some 'norms'. The method of working downwards, and dividing up a continent, country or realm can lead one to as many regions as are necessary to describe an area at a given rank. The geographer's interest may in fact be diluted over a wider area, but this method does not prevent him from concentrating his attention as certain regions reveal themselves as specially interesting and worth more attention than others. Moreover, by involving several regions at once, this method must provide comparisons and contrasts, if not absolute contrasts.

The truism that contrast is the essence of regional geography is worth repeating at times like this. Indeed, if one is describing a single, isolated region, presumably because it has special features, then this means that it is different from something else; but from what? Is the standard some theoretical norm, or a region on the Greenwich meridian, or on the equator, or the geographer's home? This presumptuous attitude of the first geographers in the field adopting their own regions as the standards for the rest of the world is well known, but hard to justify. It leads to the embarrassing situation of more exceptions than the original rule as more of the world is examined in detail, as, for example, when W. M. Davis established river erosion as the 'normal' type of erosion. The only safe and correct way, because a norm or standard cannot exist, is to compare and contrast at least two regions. Dealing with one region, having a local interest, and lacking comparisons, one may well adopt a different attitude to the region from that which might develop if one were dealing with many regions, had a wider interest, and were provided with comparisons by the very nature of the work. In fact not only does dealing with several regions provide comparisons and contrasts, it provides or reveals themes, and the very contrasts help to crystallise out the themes which epitomise the regions.

As a final statement on approach and method, the main points are tabulated here:

DIRECT OBSERVATION	INDIRECT OBSERVATION
Working out from one point	*Working downwards, division*
1. Microgeography	1. Macrogeography
2. Amateur work, useful for learning; or intensive original professional research	2. Professional work, for analysis and teaching
3. Artificial or external standards	3. Own standards by contrast
4. Concentrated interest	4. Interest in many parts of the total area, but not necessarily equally in every part, not exhaustive
5. Importance of slight variations in physical phenomena, slope, microclimate, types of grass	5. Contrast of major landforms, climates, soils, forest, grass, desert
Importance of variations within one culture, farm methods, building materials, landscape, dialect	Complete unimportance of these trivia. Major generalisations and contrasts
6. Facts not available in readily usable form. Some direct observation essential, to collect facts for small areas	6. Facts published by political and administrative units. One uses the results of others' work

7. Choose centre	7. Choose realm
8. *Map the twenty topics, plus individual crops, industries, etc.*	
9. Work outwards until the topics have changed beyond the established criteria	9. Divide each map at breaks in continua or at significant values
10. Correlate twofold to twenty-fold in the hope of some coincidence of boundaries	10. Correlate twofold, threefold, etc., which will give diminishing correlation
11. Characterise the region by the features at the core. If necessary, define by the boundary of one topic	11. Discard unimportant topics, add others as necessary. When boundaries are no longer of help, concentrate on the cores revealed in the process. The wide boundary zones are of no value, they simply isolate the cores by a process of elimination

Hartshorne points out that while world-division is the more attractive method, it requires world-knowledge first. Thus many world geographies *present* the information by dividing the whole, but the information was *collected* by working upwards and building the units. Working upwards and downwards are not mutually exclusive and can be used together in research and presentation.

Either at the beginning or the end, it is essential to make clear:

1. The Rank. The scope and degree of generalisation.
2. The criteria used to define each topic and each region.
3. The reliability and sources of the information.
4. The dates of the information and the study.
5. The purpose of the study.

[1] Odum and Moore, op. cit.

[2] Linton, op. cit.

[3] Ibid.

[4] Whittlesey, D., p. 19 in *American Geography: Inventory and Prospect*, eds. P. E. James and C. F. Jones, Syracuse University Press, New York, 1954

[5] My italics

[6] Dury, G. H., *The British Isles*, pp. 289–90, has overlapping regions of farming and industry in Lancashire

[7] Whittlesey, D., 'Southern Rhodesia, an African compage', *A.A.A.G.*, *46*, 1956

[8] Letters 18 inches high painted on a house in Athlone, June 1955

[9] Whittlesey, *American Geography*, p. 45. See fn. 4 above

[10] In practice information is coming in in haphazard order once the work is started

[11] For a compage described by an historian, not hidebound by geographic tradition, see *History of the Conquest of Peru*, W. H. Prescott, Everyman, chs. 4 and 5. Humboldt did the research, and Prescott the writing

[12] Stamp, L. D., *Applied Geography*, Pelican, 1960

[13] Freeman, T. W., *Geography and Planning*, Hutchinson University Library, 1958

[14] Jackson, J. N., *Surveys for Town and Country Planning*, Hutchinson, 3rd edn., 1967

[15] Houston, J. M., *The Western Mediterranean World*, Longmans, 1964

[16] Fawcett, C. B., *Provinces of England*, Hutchinson, 2nd edn., 1961

10

CONCLUSIONS

As suggested in Chapter 5 on regionalism, the purpose of a regional description is not always clear, or not always what it would seem to be. Again this is in large part the result of geographers being so concerned to get on with the job that very few give any indication that they are conscious of any purpose but description. In fact one is so used to regional geography being innocently academic that the few examples of regionalism are that much more dangerous. Therefore Whittlesey insisted that the geographer should examine *and state* his motives in the same way that he should examine and clarify many other hazy assumptions about other aspects of the work. He grouped the purposes of regional study under three headings, academic study for its own sake, practical study to help in planning, and prejudiced study to bring about some preconceived end; but there are several familiar purposes which do not properly fit under any of these headings but which all exhibit a partiality or bias.

1. *Partial study*

To study the influence of environment on Man.
To search for Zusammenhang.
To illustrate a preconceived theme or to put forward a point of view.

Such study is different from the academic study of the compage, which does limit its aims, but does not confine itself to seeing only one type of cause and effect, and ignoring the rest, as the studies above do.

2. *Prejudiced study*

This is specifically an attempt to change the region to some ideal, and so the danger exists of ignoring or concealing facts which may be major obstacles, in the enthusiasm and fervour of bringing about this ideal.

3. *Practical study*

Practical in the sense of useful applications being made.

To provide information for planners, characterised by the attitude of examining all the facts in order to see what ends are possible, rather than choosing an end and making the facts fit. Economics have ruined this type of study. The geographer making the practical study says, in effect, if A is wanted, then account must be taken of XY and Z; but the planners then take account only of X and Y, because Z is too expensive—or a social rather than a material consideration.

4. *Academic study*

(i) For systematic description.
To study all areal relationships.
To satisfy one's curiosity.

(ii) To reveal the regional character.
To contrast neighbouring regions.
To compare distant regions.

The purposes of academic study in group (i) probably are impossible to achieve by one man in his lifetime, and if these were the common purposes of most regional geography, Kimble's criticism would be perfectly justified. In fact, as has been shown, the purposes in group (ii) are the ones commonly found in practice. To a certain extent they are partial, in being concerned with only certain aspects of totality, but there is no prejudice and no ulterior practical motive. The very first purpose mentioned above, 'to study the influence of the physical environment on Man', is quite common in some types of geography, for the belief in determinism is quite strong. Bias and the personal point of view are advocated in this chapter, but a bias and a theme which develop from the facts, not which are imposed *a priori*. Thus it seems to the writer that there is an essential difference between the attitude 'let us look at this region *for* . . .' and 'let us look at this region to see what is actually there'. Even the most devoted possibilist may in fact come across one region where even he has to admit that man has been controlled mainly by his environment, but this admission will result from specific, concrete facts. Having the facts, any geographer may then go on to point up contrasts which appeal to him, or to compare two widely separated areas where similarities had not been noticed before, or to select significant topics as the key to the nature of a region. If we agree with Vidal de la Blache that regions are medals struck in the likeness of a people, then we must study the raw material for the medal; the motives for striking it; and the physiognomy of the people. Only then can we decide whether one leg of the tripod is more important than the other two; whether landscape or Man has been dominant.

In a sense, regional geography, and especially the description of the compage, must be pragmatic. The geographer must not have such rigid aims or ideals that he is unable to adapt to the situation, unable to adjust to facts of the earth's surface as they are, rather than as he would like them to be. Compage description demands the best methods, motives and minds, and unfortunately reveals the lack of these in some of our geography. The individual geographer is the most important factor in the work. This at once isolates geography from natural science on another count, for the essence of the natural sciences is that the observer shall have no influence on what is observed. Any experiments to try to disprove[1] natural laws must be so contrived that they can be repeated by other observers with exactly the same results. In contrast, each compage description is, or should be, unique, even if the same area has been studied at the same time. Perhaps this is one answer to Kimble's criticism that one man cannot describe a region. Rather than share out the work between a geologist, climatologist, economist, etc., the combined result may be achieved by several regional geographers each giving his partial view of reality. The result may be achieved this way, but there is no suggestion here that it will, should or must be.

The argument as to whether one man can give a complete regional description or not can be decided simply by stating the purpose of the description. If the purpose is to compile a complete, exhaustive, systematic account of every detail of every topic in one part of the earth's surface, then an army of workers is needed to produce something like the Admiralty Handbooks which many people use as sources of facts, but nobody ever reads for diversion. The end justifies the means, the end is all-important, and not one comma can be out of place. The ideal purpose for regional geography, however, is selfishness. If the purpose is to satisfy the geographer's own interest, if it arises from his anxiety to know about certain parts of the world, then it must be done by one man. The means justify the end, however short, selective and unusually arranged the final description may be. The student who wants facts from the resultant book may be infuriated, but the more it is inspired with the geographer's interest and compulsion to communicate, the more readable and enjoyable it will be.

The line must be drawn between such works as the official history of the war, or the land-use survey, and works such as Bullock's study of Hitler[2] or Siegfried's of the Mediterranean.[3] In the first two the collection and publication of all the possible facts is the overriding consideration, while in the last two one becomes aware of and shares the author's absorption in his subject. The end product is still vitally important, but not at the expense of the proper means by which it was achieved. Making the sole purpose of regional geography a flawless, characterless, mass-produced reference book ignores the main function of geography. The geographer is selfish in indulging his own interest in a region, but in a well-written description where the geographer is not ashamed to reveal his interests, the reader may be able to share these interests and go along with the development of his ideas, although this is much more common when reading history. When Kimble, Whittlesey and others suggest that regional descriptions ought to be written by a committee, they are putting perfection and economy above people. There are many suggestions along similar lines in other spheres of life, and they are all frightening.

Certainly machines can think more quickly, and work more tirelessly than people, but if this slick efficiency were carried to extremes, we would all feed from coin-in-the-slot machines, live in air-conditioned, centrally heated, artificially lighted tenements like battery hens, have all our work done for us; and be bored to death. While most of us need a purpose in life, the worst thing that can happen to us is to achieve that purpose, for then there is nothing left. Because the purpose is so important, we constantly search for a quicker or easier method of achieving it, as in the case of suggesting that committees get this geography written once and for all, as soon as possible. But we realise in the end that equally important is the function of working towards the goal, and that more satisfaction is gained from steady progress than final success. Even the time, thought and energy expended in trying to avoid hard work is more satisfying than actually avoiding it for very long.

The attitude that efficiency and economy are more important than people is Totalitarian. The urge to group work is Communist. Therefore these suggestions are all the more surprising and alarming in that they are repeated and endorsed by a group of American geographers. This kind of paradox is familiar in many people who publicly advertise their theories, then in practice do what is expedient, and end up by contradicting all their original beliefs. The American solution to the efficiency drive is for regional geography to be written by a team of fifteen specialists, one for each of the common topics in the compage. There is a feeble suggestion at the end that one geographer, poor fellow, might co-ordinate the work of the fifteen. Yet the result would still be fifteen separate monographs and the geographer-cum-editor would need the editor's usual disclaimer at the beginning of the book that he did not necessarily agree with the views of his contributors. Though few professionals would ever admit it, the Americans are here tottering on the brink of the revelation that geography is not just practical fieldwork. One man cannot collect all the facts, but one man, or close colleagues, must analyse them. Therefore, the geographer can be purely academic, providing information has been published on every topic in which he is interested, and the final work must be done inside one brain, inside the brain of one man.

The individual regional geographer, then, has two inalienable functions, to analyse and dramatise the mass of collected facts. He must dramatise in two senses, first in the sense of concentrating all the interesting points in the foreground, and secondly in the sense of giving a much wider view than is genuinely realistic. Here the geographer has much more in common with the historian, the novelist, the playwright and the painter than with any geologist or economist. The condensation in history, novels and plays is an accepted contrivance which is just as much a feature of geography. The wider view, however, is not as easy to conceive. The historian may well write about a period much longer than any man's lifetime, but the novelist who attempts this often causes his reader to forget the beginning by the time he reaches the end, and the total, all-embracing view may never be achieved. Some novelists are well known for setting their stories in definite regions, and for conveying the regional atmosphere in their works, but the two aspects of dramatisation suggested here, emphasis of the salient points of a very large area, are easiest to see in the short, simple, racy thrillers by such writers as Nicholas Freeling and Stephen Ransome.[4] From the geographical point of view such books are easy to analyse, and one realises that Freeling or Ransome create an exciting background not by sketching an imaginary region, as does Ed McBain, but by selecting, concentrating and emphasising a few interesting landmarks out of a huge area. Ransome can make Florida, or the Eastern seaboard of the USA, an exciting maze of marinas, docks, motels, market gardens, light engineering firms, bus stations, and so on. Freeling can do the same for Holland and Belgium, giving the flavour of everyday Europe by his juxtaposition of an Amsterdam street, a village shop in the Geest, the new factory, the bustling port of Rotterdam, the Belgian frontier post and the lonely hut of the Rijkwaterstaadt. The trick becomes obvious, one simply never mentions hundreds of square miles of humdrum countryside. Not only does one concentrate on the fascinating features within the field of view, but one also extends that field of view well over the horizon to gather together more interesting details. Anyone sufficiently impressed by these descriptions to go to the real places in question would probably be bitterly disappointed to find that the landscapes of Florida or Holland are no more detailed, and no more interesting, than the landscape round his own home. They are simply different, and amenable to the attentions of the writer, the painter and the geographer. The final description is simply more vivid than the view at any single point on the ground. If adventure consists of reading the results of someone else's extreme discomfort, regional geography consists of reading the results of someone's prolonged analysis and condensation of the commonplace.

The regional geographer assembles all the facts which may be necessary, then he thinks about them, and thinking is not a group activity. The facts are digested, analysed and re-synthesised into a different form, which is the geographer's personal view of reality tempered by his training. Therefore, if the task *is* too big for one man, then in the opinion of the writer the answer is not to confine each man to a single topic, and make him a general geographer, a specialist unable to communicate with other specialists, but to confine each man to a region small enough for him to comprehend, to make him a regional geographer, a complete man among men.

The compage

This word, and many of the comments in the early chapters, suggest that the compage is something different from the region. If this suggestion has been very strong, then the present chapter may have caused some confusion which must be sorted out. The real confusion arises from the many common uses of the word 'region'. The word *is* so common that it is a waste of time to try to give it more precise meaning, or to try to force people to use it in a more precise manner. The word 'compage', therefore, has been adopted to refer to one particular aspect of regional geography, to one particular type of region which is already quite familiar (although not under the name of compage) and, most of all, to one particular method of working. The compage type of region and regional method have been discussed at some length, but the following table summarises some of the more important points, and emphasises certain essential contrasts.

Traditional region	*Compage*
1. Importance of the framework of relief	1. Importance of the main topic at the centre
2. Emphasis on area, shape, size and position	2. Emphasis on the core and character
3. Implication that boundaries of all topics coincide	3. Definition by the boundary of one topic
4. Rigid list of topics in number and order	4. Variation of topics in number, order and emphasis
5. Encyclopaedic, and therefore not always inspired	5. Selective, and presented because of some genuine interest
6. Preoccupation with the physical setting	6. Concern with man at the centre

Too often, regional geography is concerned with the region only as a convenient method for dividing up areas. It should now be clear, however, that one's attention should be directed to a consideration of the contents of the regional area, the arrangement of these contents, and the differential emphasis in the final presentation. The compage approach, in contrast to the traditional approach, is more likely to direct attention to these other three features.

Finally, it must be admitted that the compage method is by no means perfect; but it does seem to the writer to be the best yet devised. In his search for a method of regional division, Hartshorne implied something very similar to the compage long before Whittlesey, but made it clear that he wished for something better.[5] Hartshorne was looking for a method which could be applied easily, was not repetitive, and which would be productive of useful concepts. He was quite prepared for the method to be arbitrary, subjective, and variable (essential characteristics of the compage method) providing it served its purpose. Therefore he discarded many well-known systems, e.g.

Ritter and Hettner	Groups of specific regions
Herbertson	Natural (physical) regions, in fact based on climate
Van Valkenburg, Stamp, Unstead	Similar to Herbertson
James and Passarge	Geographic regions, in fact based on vegetation,

because it seemed to him that a division by means of the same topic for the whole world would not work. Unfortunately, division using now one topic, now another, seemed so awkward to Hartshorne that he passed over this method. However, he was not even satisfied to divide the world by a combination of topics because this would not result in generic types of regions or produce useful general concepts. The greatest danger in all these methods is the assumption that the regions of any one topic fit perfectly the regions of other topics. The danger is greatest in Herbertson's natural regions, where the assumption was that the units of relief, climate, soil and vegetation coincide. Any critical examination of Herbertson reveals, however, that his natural regions are in fact climatic regions, and the relief, soil and vegetation regions were assumed to fit. Such systems could be widely used only as long as people were prepared to believe that

(a) natural vegetation exists,
(b) natural vegetation is the resultant of climate, soil and relief,
(c) Man adapts himself completely to his physical environment.

It is now clear that this is not the case over much of the earth's surface. Moreover, one topic is not sufficiently important to be taken as the same starting point for regional description everywhere in the world. While this may not have been intended, Hartshorne points out that most of the time this method of always starting with the natural region, or physical framework, leads to determinist ways of thinking. Having started everywhere, every time, with a thorough description of the natural environment, too many geographers have fallen into the error of going on to explain human activity by reference to the natural environment alone. Regional geography has been most partial and one-sided. In fact, even when political, economic, social and psychological factors have been admitted as more important than physical factors, geographers have ignored them on the ground that these factors are not the proper concern of geography. The compage approach is essentially different from this, because while it may be necessary to define a given region by the criterion of one topic, other topics will be used for other regions, and there is never any assumption that everything in the region must be explained by the topic which is simply used to characterise it, or to define it.

In his search, Hartshorne considered possible systems of cultural regions based on such things as political units, groups of population, and different economic activities. Jones and Whittlesey on one hand, and Hartshorne and Dicken on the other, attempted to establish element-complexes as a basis for regional division,[6] but there proved to be at least two serious snags. First, as the study becomes more detailed and more complicated, so many topics which vary independently have to be brought in, e.g. crop combinations, method of cultivation, field patterns, land tenure, farm size, rural settlement and intensity of farming, that in the end one is left with specific regions and no types. Second, even if some of these human features are generalised, as the boundaries of the five topics of physical geography do not correspond, it is hardly surprising that very rarely are the boundaries of the topics of human geography found to correspond with those of physical geography.

On the other side of the balance there is one decided advantage in starting with a framework of cultural regions and in starting always with a complex (multiple-topic region in Whittlesey's later nomenclature). This is that one is much more likely to avoid false determinism and to avoid mistaking climatic or relief regions for full 'natural regions'. Starting with economic activities, one might study the extent to which the natural environment is an obstacle to Man; thus suggesting that Man decides his aims for himself. Moreover, in describing and explaining a complex cultural region one is forced to take into account *all* the factors which have produced the present surface phenomena. The only sound method of working is to proceed from the known to the unknown. Throughout the world the cultural landscape is self-evident but, in spite of what the school textbooks and atlases confidently show, the facts of the geological structure, detailed relief, microclimate, soil profiles and wild vegetation have not been studied or mapped in any detail except in Europe.

Hartshorne's conclusions, then, agree with the conclusion that the compage method is the most suitable and realistic yet produced, despite its obvious shortcomings. He concludes that we must have as many regional systems as are necessary, based on individual topics, or on simple combinations to suit the areas under study. His preference is for a system 'based on the great number of cultural features synthesised by Man in his productive use of the land'.[7] His argument, which is developed in convincing detail, reveals the simple logic that

1. there are few self-evident boundaries in nature,
2. imposing one system of arbitrary boundaries will give as many false regions as true ones,
3. as each region has a unique place in relation to all others, generic classifications are out; regions are specific,
4. and any world system of specific regions will be possible only when all the geography has been written; therefore no perfect world system can be available as a starting point.

The hypothetical boundaries we adopt as a working basis may be changed out of all recognition by the time we have finished.

To reverse the first analogy in this book, while we are as familiar with atlas maps as we are with regional descriptions, too often we ignore the tiny print in one corner which says Equal-Area, Equidistant, Orthomorphic or Transverse Mercator. Yet this tiny print is there as a reminder that the map is only a device to try to show the surface of a sphere on a flat surface. A better method would be the exclusive use of globes, but atlases are much cheaper and much more convenient. Many people ignore the reminder in the corner of the map; others look for it at once in order to be quite sure of the shortcomings and qualities of that particular map. It is exactly the same case with regional descriptions. The regional division is a device which we find so convenient that its use is preferable to other methods. At the same time, each regional geographer should do much more than put a tiny note in the corner; he should emphasise the purpose of his method, and its artificial nature.

While this regional device necessitates division and the use of some form of boundaries, the pursuit of more universal boundaries or greater precision seems to be a wild-goose chase. What is needed is greater precision in thought, rather than in drawing lines. This difficulty in finding boundaries has combined with the difficulty of analysing all the material relevant to regional description in causing some authorities to conclude that complete geographical description is too much for one individual. If the work must be limited, then the present writer would insist most strongly that for geography to continue, the individual should limit himself or herself to a smaller area rather than to one topic. The essence of geography is the infinite variety and mixture of phenomena on the earth's surface. Those who specialise in the general geography of one topic may be geologists, geomorphologists, economists or demographers, but they are not, in the strict sense of the term, geographers.

Limitation of each person to one topic results from the concept of geography as a practical task, as fieldwork. One may regard geographical research as an imposition, and a piece of work to be completed once and for all as soon as possible—like winning the war so that one may return to normal peacetime life. If this is the case, then the more experts, money and equipment which can be concentrated on the problem the better. This view is valid in some ways; the Admiralty Handbooks were compiled in this way as part of our intelligence service during the war. At present accurate, up-to-date facts about Africa, India, Pakistan and South-east Asia must be compiled rapidly if Europe and North America are to help develop these countries wisely and soundly. But this type of applied research[8] must not be mistaken for academic geography.

The motive for geography is interest, and the prime purpose of the work is to satisfy that interest. Something attracts a geographer to a continent, country or area. Something, anything, attracts him and makes him find out to satisfy his own curiosity. There is the anxiety to know, and then compulsion to tell. This academic interest is the spirit and purpose of geography.

We come to the conclusion that there is no one definition of a region, and no one method of recognising, delimiting or describing a region. Just as the world is unique; just as each continent and country is unique; so each region is unique and requires a unique approach. In the days when natural science tries to make us slaves of a system, clinging to systematisation for comfort, we reach the fearsome conclusion that for this discipline, at least, we cannot fall back on some ready-made machinery—like some bureaucrat reaching for his book of rules. In the natural sciences a few individuals, as ever, have to think completely for themselves, originate ideas, rely on their own judgment; but the bulk of modern research, growing like yeast, is in fact routine work. In contrast, regional geography requires the long-praised virtue (still praised in public but usually disregarded in private practice) of every geographer, however limited and insignificant his region, judging an original case on its merits. This essential nature of the region should be proclaimed all the time. The attempts to show that geography is a natural science and the attempts to produce hierarchies, classifications and the little observer's guide to types of regions have continued much too long. If we shake off this embarrassing disguise geography will lose all those who need the order, orthodoxy and safety of a natural science to which to cling, and will be infinitely better for the loss. There is no order and orthodoxy in the features we study, and there should be the minimum in the study itself. When a country is to be divided into regions for the two sound reasons that (a) divisions suggest themselves and (b) division is necessary for comprehension, then there is no reason on earth why (a) some earlier division by some other person should be followed and regarded as sacrosanct, (b) why a method of division used in Manchuria should be applied to Salford.

These unique regions, each requiring a completely fresh approach, perhaps a unique choice of boundary, and certainly a unique description, are not material for the unimaginative hack, nor are they amenable to modern methods of mass data-processing by which the North Americans in particular seem to hope to reduce Man's knowledge to a punched tape or a teaching machine. They require genuine interest for their own sake, imagination, and an original approach to get this excitement across to the reader vividly, and—most challenging of all—they require the geographer to see, judge, measure and decide for himself.

[1] This *is* the modern attitude towards experiments

[2] Bullock, Alan, *Hitler, a Study in Tyranny*, Odhams, 1952

[3] Siegfried, André, *The Mediterranean*, Cape, 1948

[4] Both published by Gollancz

[5] Hartshorne, op. cit., ch. 9

[6] Ibid., p. 340

[7] Ibid., p. 346

[8] Usually called 'applied geography' in Britain

BIBLIOGRAPHY

(*A.A.A.G.: Annals of the Association of American Geographers*)

1. *Theory of regions*

CHISHOLM, M., *Rural Settlement and Land Use*, Hutchinson, 1962

COLE, J. B., *Geography of World Affairs*, Penguin, 1959

CRESSEY, G., pp. 32–3 and the several maps of regional divisions in *Asia's Lands and Peoples*, McGraw Hill, 1955

DRYER, C. R., 'Natural Economic Regions', *A.A.A.G.*, *5*, 1915

DURY, G. H., ch. 11, part 2, pp. 163–6 in *The British Isles*, Heinemann, 3rd edn., 1965

EAST, W. G., *The Geography Behind History*, Nelson, 1965

FREEMAN, T. W., *A Hundred Years of Geography*, Duckworth, 1961

GILBERT, E. W., 'Geography and Regionalism' in *Geography in the 20th Century*, ed. G. Taylor, Methuen, 3rd edn., 1960

GILBERT, E. W., 'The Idea of the Region', *Geography*, V, 45, 1960

HALL, R. B., 'The Geographic Region', *A.A.A.G.*, *25*, 1935

HARTSHORNE, R., 'The Nature of Geography', *A.A.A.G.*, *29*, 1939

HARTSHORNE, R., *Perspective on the Nature of Geography*, Murray, 1959

HARTSHORNE, R., and DICKEN, S. N., 'A Classification of the Agricultural Regions of Europe and North America on a Uniform Statistical Basis', *A.A.A.G.*, *15*, 1935

HERBERTSON, A. J., 'The Major Natural Regions', *Geographical Journal*, *25*, 1905

JAMES, P. E., 'Toward a Further Understanding of The Regional Concept', *A.A.A.G.*, *42*, 1952

JENSEN, M. (ed.), *Regionalism in America*, Madison, Wisconsin, 1951

KIMBLE, G. H. T., 'The inadequacy of the regional concept' in *London Essays in Geography*, eds. Stamp and Wooldridge, Longmans, 1951

LINTON, D. L., 'The delimitation of morphological regions' in *London Essays in Geography*, eds. Stamp and Wooldridge, Longmans, 1951

ODUM, H. W., and MOORE, H. E., *American Regionalism*, Henry Holt & Co.. New York, 1938

PARKER, W. H., Introduction to part II, p. 114, in *Anglo-America*, ULP, 1962

PHILBRICK, A. K., 'Principles of Areal Functional Organisation in Regional Human Geography', *Econ. Geog.*, *33*, 1957

PLATT, R. S., 'A Review of Regional Geography', *A.A.A.G.*, *47*, 1957

ROBINSON, G. W. S., 'The Geographical Region: Form and Function', *Scottish Geographical Magazine*, *69*, 1953

ROXBY, P. M., 'The Theory of Natural Regions', *Geography*, *13*, 1925–6

SHABAD, T., pp. xi–xvi on methodology in *Geography of the U.S.S.R.*, Columbia Univ. Press, New York, and OUP 1951

SPATE, O. H. K., ch. XIII, pp. 351–65 in *India and Pakistan*, Methuen, 3rd edn., 1967

ULLMAN, E. L., 'Rivers as Regional Bonds', *Geog. Review.*, *41*, New York, 1951

UNSTEAD, J. F., 'A System of Regional Geography', *Geography*, *18*, 1933

UNSTEAD, J. F., 'Classifications of Regions of the World', *Geography*, *22*, 1937

UNSTEAD, J. F., Glossary and Index to Regional Terms in *The British Isles*, ULP, 6th edn., 1964

VANCE, R. B., *Human Geography of the South*, Chapel Hill, N. Carolina, 1932, Introduction

WHITTLESEY, D., 'Major Agricultural Regions of the Earth', *A.A.A.G.*, *26*, 1936

WHITTLESEY, D., 'The Regional Concept and the Regional Method', in, *American Geography: Inventory and Prospect*, eds. James and Jones, 1954

WOOLDRIDGE, S. W., and EAST, W. G., *The Spirit and Purpose of Geographys* Hutchinson, 2nd edn., 1960

2. *Regional texts*

The regional geography under consideration throughout this work is that presented to the sixth-former, the undergraduate and the general public. In contrast to some other works in section one, therefore, the regional descriptions referred to are those currently available as text books. The following list includes most of the leading texts in use in Britain and known to the writer. They are quoted regardless of their merits or shortcomings, as they form the background against which many generalisations have been made.

ATWOOD, W. W., *The Physiographic Provinces of North America*, Ginn, Boston, 1940

BARBOUR, K. M., *The Republic of the Sudan*, ULP, 1961

Berg, L. S., *Natural Regions of the U.S.S.R.*, Macmillan, 1950
Boateng, E. A., *A Geography of Ghana*, CUP, 1959
Bowen, E. G. (ed.), *Wales*, Methuen, 1957
Cole, J. P., *Italy*, Chatto and Windus, 1964
Cole, M. M., *South Africa*, Methuen, 1960
Cressey, G. B., *Asia's Lands and Peoples*, McGraw Hill, 2nd edn., 1955
Cumberland, K. B., *South-west Pacific*, Methuen, 2nd edn., 1958
Daysh, G. H. J., *Studies in Regional Planning*, Philip, 1951
Demangeon, A., *The British Isles*, Heinemann, 1939
De Martonne, E., and Demangeon, A., *La France, Géographie Universelle*, Tome VI, Armand Colin, 1947
De Martonne, E., *Geographical Regions of France*, Heinemann, 1933
Dickinson, R. E., *Germany*, Methuen, 2nd edn., 1964
Dobby, E. H. G., *South East Asia*, ULP, 8th edn., 1966
Dobby, E. H. G., *Monsoon Asia*, ULP, 3rd edn., 1966
Dury, G. H., *The British Isles*, Heinemann, 3rd edn., 1965
Dury, G. H., *The East Midlands and the Peak*, Nelson, 1963
Fisher, C. A., *Southeast Asia*, Methuen, 1964
Fisher, W. B., *The Middle East*, Methuen, 5th edn., 1962
Fitzgerald, W., *Africa*, Methuen, 9th edn., 1964
Freeman, T. W., *Ireland*, Methuen, 3rd edn., 1965
Freeman, T. W., and Rodgers, H. B., *Lancashire, Cheshire and the Isle of Man*, Nelson, 1967
Garland, J. H., *The North American Midwest*, Chapman and Hall, 1955
Hoffman, G. H. (ed.), *A Geography of Europe*, Methuen, 2nd edn., 1961
Houston, J. M., *The Western Mediterranean World*, Longmans, 1964
James, P. E., *Latin America*, Cassell, 3rd edn., 1959
Jones, L. R., and Bryan, P. W., *North America*, Methuen, 10th edn., 1963
Monkhouse, F. J., *A Regional Geography of Western Europe*, Longmans, 2nd edn., 1964
Mutton, A. F. A., *Central Europe*, Longmans, 1961
Newbigin, M. I., *Southern Europe*, Methuen, revised edn. 1952
O'Dell, A. C., and Houston, J. M., *Central Scotland*, Philip, 1951
O'Dell, A. C., and Walton, K., *The Highlands and Islands of Scotland*, Nelson, 1962
Ogilvie, A. G. (ed.), *Great Britain: Essays in Regional Geography*, CUP, 1928
Ormsby, H., *France*, Methuen, 3rd edn., 1964
Platt, R. S., *Latin America: Countrysides and United Regions*, McGraw Hill, New York, 1943
Putnam, D. F., *Canadian Regions*, Dent, 1952
Robequain, C., *Malaya, Indonesia, Borneo and the Philippines*, Longmans, 2nd edn., 1958
Robinson, K. W., *Australia, New Zealand and S.W. Pacific*, ULP, 1960
Shabad, T., *Geography of the U.S.S.R.*, OUP, 1951

SHACKLETON, M. R., *Europe*, Longmans, 7th edn. revised, 1964

SHANAHAN, E., *South America*, Methuen, 11th edn., 1963

SMAILES, A. E., *North England*, Nelson, 1960

SPATE, O. H. K., *India and Pakistan*, Methuen, 3rd edn., 1967

STAMP, L. D., *Asia, An Economic and Regional Geography*, Methuen, 10th edn., 1959

TAYLOR, G., *Australia*, Methuen, 7th edn., 1961

TREWARTHA, G. T., *Japan*, Methuen, 2nd edn., 1965

UNSTEAD, J. F., *The British Isles*, ULP, 6th edn., 1964

UNSTEAD, J. F., *A World Survey, A Systematic Regional Geography*, ULP, 6th edn., 1965

WALKER, D. S., *A Geography of Italy*, Methuen, 1958

WELLINGTON, J. H., *Southern Africa*, CUP, 1955

WHITE, C. L., and FOSCUE, E. J., *Regional Geography of Anglo-America* Prentice-Hall, rev. edn., 1965

3. *Regionalism*

Regionalism in World Economics: *Bulletin of the Institute of World Economics, 3*, 1945

DICKINSON, R. E., *City Region and Regionalism*, Kegan Paul, 1947

DICKINSON, R. E., 'Regionalism in Modern Germany', ch. 14 in *Germany*, Methuen, 2nd edn., 1964

FAWCETT, C. B., *Provinces of England*, Hutchinson University Library, 2nd edn., 1961

FREEMAN, T. W., *Geography and Planning*, Hutchinson, 3rd edn., 1967

GILBERT, E. W., 'Practical Regionalism in England and Wales', *Geog. Journal, 94*, 1939

GILBERT, E. W., 'The Boundaries of Local Government Areas', *Geog. Journal, 111*, 1948

JACKSON, J. N., *Surveys for Town and Country Planning*, Hutchinson, 1963

JENSON, M., *Regionalism in America*, Madison, Wisconsin, 1951

ODUM, H. W., and MOORE, H. E., *American Regionalism*, Henry Holt & Co., New York, 1938

SENIOR, D., *The Regional City*, Longmans, 1966

SPATE, O. H. K., The region as a work of art, in *Let Me Enjoy*, Methuen, 1966

STAMP, L. D., *Applied Geography*, Pelican, 1960

SZAVA-KOVATS, E., The present state of landscape theory and its main philosophic problems, in *Soviet Geography: Review and Translation*, American Geographical Society, New York, vol. VII, no. 7, September, 1966

YALEM, R. J., *Regionalism and World Order*, Public Affairs Press, Washington D.C., 1967

Note. R. E. Dickinson's *City Region and Regionalism* examined in Chapter 5, was the original work of 1947. This has now been completely revised and the title has been changed to *City and Region: A Geographical Interpretation*, Routledge and Kegal Paul, 1964

APPENDIX TO FIGURE 3

Examples of interstitial and atypical areas, between and inside regions respectively. See Figure 3, p. 57.

1. Whyte, W. H., units isolated by the urban plan, in *The Organisation Man*, Cape, 1957
2. McCarty, H. H., subsistence areas, in *The Geographical Basis of American Economic Life*, Harper, 1940
3. Smailes, A. E., service vacuums, in *The Geography of Towns*, Hutchinson University Library, 1953
4. Taylor, E. G. R., gaps between city-regions, see map p. 378 in *Urban Geography*, Griffith Taylor, Methuen, 1949
5. James, P. E., 'unused areas' on his maps in *Latin America*, Cassell, 1959
6. Taylor, G., gaps between regions 11, 14, 15 and others on the map on p. 130 in *Australia*, Methuen, 1940
7. Spate, O. H. K., distinguishes transitional and outlier types of areas, e.g. outlier of the Aravallis in the Punjab, in *India and Pakistan*, Methuen, 1966
8. Dury, G. H., overlapping regions, Fig. 83, p. 199, in *The British Isles*, Heinemann, 1965
9. Monkhouse, F. J., atypical areas, Ghent inside Flanders and Brussels inside Brabant, while parts of Maritime Flanders and the Ardennes are outliers, see his *Western Europe*, Longmans, 1964
10. Unstead, J. F., exclaves, see also his glossary and index to regional terms in *The British Isles*, ULP, 1964
11. Trewartha, G. T., see his detailed subdivision of Tohoku, Hokkaido and Central and Southwestern Japan in *Japan*, Methuen, 1965
12. Dickinson, R. E., gives similar detailed subdivision in *Germany*, Methuen, 1964

INDEX